Social Life in the Insect World

Insect World

J. H. Fabre

Contents

CHAPTER I THE FABLE OF THE CIGALE AND THE ANT 7

CHAPTER II THE CIGALE LEAVES ITS BURROW...22

CHAPTER III THE SONG OF THE CIGALE 31

CHAPTER IV THE CIGALE. THE EGGS AND THEIR HATCHING................40

CHAPTER V THE MANTIS.--THE CHASE..55

CHAPTER VI THE MANTIS.--COURTSHIP 62

CHAPTER VII THE MANTIS.--THE NEST 67

CHAPTER VIII THE GOLDEN GARDENER.--ITS NUTRIMENT....................78

CHAPTER IX THE GOLDEN GARDENER--COURTSHIP...............................84

CHAPTER X THE FIELD-CRICKET...90

CHAPTER XI THE ITALIAN CRICKET 97

CHAPTER XII THE SISYPHUS BEETLE.--THE INSTINCT ....

    OF PATERNITY 101

CHAPTER XIII A BEE-HUNTER: THE *PHILANTHUS AVIPORUS*..............*111*

CHAPTER XIV THE GREAT PEACOCK, OR EMPEROR MOTH..................130

CHAPTER XV THE OAK EGGAR, OR BANDED MONK 145

CHAPTER XVI A TRUFFLE-HUNTER: THE *BOLBOCERAS GALLICUS* *155*

CHAPTER XVII THE ELEPHANT-BEETLE 168

CHAPTER XVIII THE PEA-WEEVIL--BRUCHUS PISI *181*

CHAPTER XIX AN INVADER.--THE HARICOT-WEEVIL............................197

CHAPTER XX THE GREY LOCUST...209

CHAPTER XXI THE PINE-CHAFER..220

NOTES 225

SOCIAL LIFE IN THE INSECT WORLD

BY

J. H. Fabre

CHAPTER I
THE FABLE OF THE CIGALE AND THE ANT

Fame is the daughter of Legend. In the world of creatures, as in the world of men, the story precedes and outlives history. There are many instances of the fact that if an insect attract our attention for this reason or that, it is given a place in those legends of the people whose last care is truth.

For example, who is there that does not, at least by hearsay, know the Cigale? Where in the entomological world shall we find a more famous reputation? Her fame as an impassioned singer, careless of the future, was the subject of our earliest lessons in repetition. In short, easily remembered lines of verse, we learned how she was destitute when the winter winds arrived, and how she went begging for food to the Ant, her neighbour. A poor welcome she received, the would-be borrower!--a welcome that has become proverbial, and her chief title to celebrity. The petty malice of the two short lines--

Vous chantiez! j'en suis bien aise,
Eh bien, dansez maintenant!

has done more to immortalise the insect than her skill as a musician. "You sang! I am very glad to hear it! Now you can dance!" The words lodge in the childish memory, never to be forgotten. To most Englishmen--to most Frenchmen even--the song of the Cigale is unknown, for she dwells in the country of the olive-tree; but we all know of the treatment she received at the hands of the Ant. On such trifles does Fame depend! A legend of very dubious value, its moral as bad as its natural history; a nurse's tale whose only merit is its brevity; such is the basis of a reputation which will survive the wreck of centuries no less surely than the tale of Puss-in-Boots and of Little Red Riding-Hood.

The child is the best guardian of tradition, the great conservative. Custom and tradition become indestructible when confided to the archives of his memory. To the child we owe the celebrity of the Cigale, of whose misfortunes he has babbled during his first lessons in recitation. It is he who will preserve for future generations the absurd nonsense of which the body of the fable is constructed; the Cigale will always be hungry when the cold comes, although there were never Cigales in winter; she will always beg alms in the shape of a few grains of wheat, a diet absolutely incompatible with her delicate capillary "tongue"; and in desperation she will hunt for flies and grubs, although she never eats.

Whom shall we hold responsible for these strange mistakes? La Fontaine, who in most of his fables charms us with his exquisite fineness of observation, has here been ill-inspired. His earlier subjects he knew down to the ground: the Fox, the Wolf, the Cat, the Stag, the Crow, the Rat, the Ferret, and so many others, whose actions and manners he describes with a delightful precision of detail. These are inhabitants of his own country; neighbours, fellow-parishioners. Their life, private and public, is lived under his eyes; but the Cigale is a stranger to the haunts of Jack Rabbit. La Fontaine had never seen nor heard her. For him the celebrated songstress was certainly a grasshopper.

Grandville, whose pencil rivals the author's pen, has fallen into the same error. In his illustration to the fable we see the Ant dressed like a busy housewife. On her threshold, beside her full sacks of wheat, she disdainfully turns her back upon the would-be borrower, who holds out her claw--pardon, her hand. With a wide coachman's hat, a guitar under her arm, and a skirt wrapped about her knees by the gale, there stands the second personage of the fable, the perfect portrait of a grasshopper. Grandville knew no more than La Fontaine of the true Cigale; he has beautifully expressed the general confusion.

But La Fontaine, in this abbreviated history, is only the echo of another fabulist. The legend of the Cigale and the cold welcome of the Ant is as old as selfishness: as old as the world. The children of Athens, going to school with their baskets of rush-work stuffed with figs and olives, were already repeating the story under their breath, as a lesson to be repeated to the teacher. "In winter," they used to say, "the Ants were putting their damp food to dry in the sun. There came a starving Cigale to beg from them. She begged for a few grains. The greedy misers replied: 'You sang in the summer, now dance in the winter.'" This, although somewhat more arid, is precisely La Fontaine's story, and is contrary to the facts.

Yet the story comes to us from Greece, which is, like the South of France, the home of the olive-tree and the Cigale. Was AEsop really its author, as tradition would have it? It is doubtful, and by no means a matter of importance; at all events, the author was a Greek, and a compatriot of the Cigale, which must have been perfectly familiar to him. There is not a single peasant in my village so blind as to be unaware of the total absence of Cigales in winter; and every tiller of the soil, every gardener, is familiar with the first phase of the insect, the larva, which his spade is perpetually discovering when he banks up the olives at the approach of the cold weather, and he knows, having seen it a thousand times by the edge of the country paths, how in summer this larva issues from the earth from a little round well of its own making; how it climbs a twig or a stem of grass, turns upon its back, climbs out of its skin, drier now than parchment, and becomes the Cigale; a creature of a fresh grass-green colour which is rapidly replaced by brown.

We cannot suppose that the Greek peasant was so much less intelligent than the Provencal that he can have failed to see what the least observant must have noticed. He knew what my rustic neighbours know so well. The scribe, whoever he may have been, who was responsible for the fable was in the best possible circumstances for correct knowledge of the subject. Whence, then, arose the errors of his tale?

Less excusably than La Fontaine, the Greek fabulist wrote of the Cigale of the books, instead of interrogating the living Cigale, whose cymbals were resounding on every side; careless of the real, he followed tradition. He himself echoed a more ancient narrative; he repeated some legend that had reached him from India, the venerable mother of civilisations. We do not know precisely what story the reed-pen of the Hindoo may have confided to writing, in order to show the perils of a life without foresight; but it is probable that the little animal drama was nearer the truth than the conversation between the Cigale and the Ant. India, the friend of animals, was incapable of such a mistake. Everything seems to suggest that the principal personage of the original fable was not the Cigale of the Midi, but some other creature, an insect if you will, whose manners corresponded to the adopted text.

Imported into Greece, after long centuries during which, on the banks of the

Indus, it made the wise reflect and the children laugh, the ancient anecdote, perhaps as old as the first piece of advice that a father of a family ever gave in respect
of economy, transmitted more or less faithfully from one memory to another, must
have suffered alteration in its details, as is the fate of all such legends, which the
passage of time adapts to the circumstance of time and place.

The Greek, not finding in his country the insect of which the Hindoo spoke,
introduced the Cigale, as in Paris, the modern Athens, the Cigale has been replaced
by the Grasshopper. The mistake was made; henceforth indelible. Entrusted as it is
to the memory of childhood, error will prevail against the truth that lies before our
eyes.

Let us seek to rehabilitate the songstress so calumniated by the fable. She is,
I grant you, an importunate neighbour. Every summer she takes up her station in
hundreds before my door, attracted thither by the verdure of two great plane-trees;
and there, from sunrise to sunset, she hammers on my brain with her strident symphony. With this deafening concert thought is impossible; the mind is in a whirl, is
seized with vertigo, unable to concentrate itself. If I have not profited by the early
morning hours the day is lost.

Ah! Creature possessed, the plague of my dwelling, which I hoped would be
so peaceful!--the Athenians, they say, used to hang you up in a little cage, the better to enjoy your song. One were well enough, during the drowsiness of digestion;
but hundreds, roaring all at once, assaulting the hearing until thought recoils--this
indeed is torture! You put forward, as excuse, your rights as the first occupant.
Before my arrival the two plane-trees were yours without reserve; it is I who have
intruded, have thrust myself into their shade. I confess it: yet muffle your cymbals,
moderate your arpeggi, for the sake of your historian! The truth rejects what the
fabulist tells us as an absurd invention. That there are sometimes dealings between
the Cigale and the Ant is perfectly correct; but these dealings are the reverse of
those described in the fable. They depend not upon the initiative of the former;
for the Cigale never required the help of others in order to make her living: on the
contrary, they are due to the Ant, the greedy exploiter of others, who fills her granaries with every edible she can find. At no time does the Cigale plead starvation at
the doors of the ant-hills, faithfully promising a return of principal and interest; the
Ant on the contrary, harassed by drought, begs of the songstress. Begs, do I say! Bor-

rowing and repayment are no part of the manners of this land-pirate. She exploits the Cigale; she impudently robs her. Let us consider this theft; a curious point of history as yet unknown.

In July, during the stifling hours of the afternoon, when the insect peoples, frantic with drought, wander hither and thither, vainly seeking to quench their thirst at the faded, exhausted flowers, the Cigale makes light of the general aridity. With her rostrum, a delicate augur, she broaches a cask of her inexhaustible store. Crouching, always singing, on the twig of a suitable shrub or bush, she perforates the firm, glossy rind, distended by the sap which the sun has matured. Plunging her proboscis into the bung-hole, she drinks deliciously, motionless, and wrapt in meditation, abandoned to the charms of syrup and of song.

Let us watch her awhile. Perhaps we shall witness unlooked-for wretchedness and want. For there are many thirsty creatures wandering hither and thither; and at last they discover the Cigale's private well, betrayed by the oozing sap upon the brink. They gather round it, at first with a certain amount of constraint, confining themselves to lapping the extravasated liquor. I have seen, crowding around the honeyed perforation, wasps, flies, earwigs, Sphinx-moths, Pompilidae, rose-chafers, and, above all, ants.

The smallest, in order to reach the well, slip under the belly of the Cigale, who kindly raises herself on her claws, leaving room for the importunate ones to pass. The larger, stamping with impatience, quickly snatch a mouthful, withdraw, take a turn on the neighbouring twigs, and then return, this time more enterprising. Envy grows keener; those who but now were cautious become turbulent and aggressive, and would willingly drive from the spring the well-sinker who has caused it to flow.

In this crowd of brigands the most aggressive are the ants. I have seen them nibbling the ends of the Cigale's claws; I have caught them tugging the ends of her wings, climbing on her back, tickling her antennae. One audacious individual so far forgot himself under my eyes as to seize her proboscis, endeavouring to extract it from the well!

Thus hustled by these dwarfs, and at the end of her patience, the giantess finally abandons the well. She flies away, throwing a jet of liquid excrement over her tormentors as she goes. But what cares the Ant for this expression of sovereign contempt? She is left in possession of the spring--only too soon exhausted when the pump is removed that made it flow. There is little left, but that little is sweet. So much to the good; she can wait for another drink, attained in the same manner, as soon as the occasion presents itself.

As we see, reality completely reverses the action described by the fable. The shameless beggar, who does not hesitate at theft, is the Ant; the industrious worker, willingly sharing her goods with the suffering, is the Cigale. Yet another detail, and the reversal of the fable is further emphasised. After five or six weeks of gaiety, the songstress falls from the tree, exhausted by the fever of life. The sun shrivels her body; the feet of the passers-by crush it. A bandit always in search of booty, the Ant discovers the remains. She divides the rich find, dissects it, and cuts it up into tiny fragments, which go to swell her stock of provisions. It is not uncommon to see a dying Cigale, whose wings are still trembling in the dust, drawn and quartered by a gang of knackers. Her body is black with them. After this instance of cannibalism the truth of the relations between the two insects is obvious.

Antiquity held the Cigale in high esteem. The Greek Beranger, Anacreon, devoted an ode to her, in which his praise of her is singularly exaggerated. "Thou art almost like unto the Gods," he says. The reasons which he has given for this apotheosis are none of the best. They consist in these three privileges: [Greek: gegenes, apathes, hanaimosarke]; born of the earth, insensible to pain, bloodless. We will not reproach the poet for these mistakes; they were then generally believed, and were perpetuated long afterwards, until the exploring eye of scientific observation was directed upon them. And in minor poetry, whose principal merit lies in rhythm and harmony, we must not look at things too closely.

Even in our days, the Provencal poets, who know the Cigale as Anacreon never did, are scarcely more careful of the truth in celebrating the insect which they have taken for their emblem. A friend of mine, an eager observer and a scrupulous realist, does not deserve this reproach. He gives me permission to take from his pigeon-holes the following Provencal poem, in which the relations between the Cigale and the Ant are expounded with all the rigour of science. I leave to him the responsibility for his poetic images and his moral reflections, blossoms unknown to my naturalist's garden; but I can swear to the truth of all he says, for it corresponds with what I see each summer on the lilac-trees of my garden.

LA CIGALO E LA FOURNIGO.

I.

Jour de Dieu, queto caud! Beu tems per la Cigalo,
  Que, trefoulido, se regalo
D'uno raisso de fio; beu tems per la meissoun.
  Dins lis erso d'or, lou segaire,
Ren plega, pitre au vent, rustico e canto gaire;
Dins soun gousie, la set estranglo la cansoun.

Tems benesi per tu. Dounc, ardit! cigaleto,
  Fai-lei brusi, ti chimbaleto,
E brandusso lou ventre a creba ti mirau.
  L'Ome enterin mando le daio,
Que vai balin-balan de longo e que dardaio
L'ulau de soun acie sus li rous espigau.

Plen d'aigo per la peiro e tampouna d'erbiho
  Lou coufie sus l'anco pendiho.
Si la peiro es au fres dins soun estui de bos,
  E se de longo es abeurado,
L'Ome barbelo au fio d'aqueli souleiado
Que fan bouli de fes la mesoulo dis os.

Tu, Cigalo, as un biais per la set: dins la rusco
  Tendro e jutouso d'uno busco,
L'aguio de toun be cabusso e cavo un pous.
  Lou siro monto per la draio.
T'amourres a la fon melicouso que raio,
E dou sourgent sucra beves lou teta-dous.

Mai pas toujour en pas. Oh! que nani; de laire,
  Vesin, vesino o barrulaire,
T'an vist cava lou pous. An set; venon, doulent,
  Te prene un degout per si tasso.
Mesfiso-te, ma bello: aqueli curo-biasso,
Umble d'abord, soun leu de gusas insoulent.

Quiston un chicouloun di ren, piei de ti resto
  Soun plus countent, ausson la testo
E volon tout: L'auran. Sis arpioun en rasteu
  Te gatihoun lou bout de l'alo.
Sus tu larjo esquinasso es un mounto-davalo;
T'aganton per lou be, li bano, lis arteu;

Tiron d'eici, d'eila. L'impacienci te gagno.
  Pst! pst! d'un giscle de pissagno
Asperges l'assemblado e quites lou rameu.
  T'en vas ben liuen de la racaio,
Que t'a rauba lou pous, e ris, e se gougaio,
E se lipo li brego enviscado de meu.

Or d'aqueli boumian abeura sens fatigo,
  Lou mai tihous es la fournigo.
Mousco, cabrian, guespo e tavan embana,
  Espeloufi de touto meno,
Costo-en-long qu'a toun pous lou soulcias ameno,
N'an pas soun testardige a te faire enana.

Per l'esquicha l'arteu, te coutiga lou mourre,
  Te pessuga lou nas, per courre
A l'oumbro du toun ventre, osco! degun la vau.
  Lou marrit-peu prend per escalo
Uno patto e te monto, ardido, sus lis alo,

E s'espasso, insoulento, e vai d'amont, d'avau.

II.

Aro veici qu'es pas de creire.
Ancian tems, nous dison li reire,
Un jour d'iver; la fam te prengue. Lou front bas
E d'escoundoun aneres veire,
Dins si grand magasin, la fournigo, eilabas.

L'endrudido au souleu secavo,
Avans de lis escoundre en cavo,
Si blad qu'avie mousi l'eigagno de la niue.
Quand eron lest lis ensacavo.
Tu survenes alor, eme de plour is iue.

Ie dises: "Fai ben fre; l'aurasso
D'un caire a l'autre me tirasso
Avanido de fam. A toun riche mouloun
Leisso-me prene per ma biasso.
Te lou rendrai segur au beu tems di meloun.

"Presto-me un pan de gran." Mai, bouto,
Se creses que l'autro t'escouto,
T'enganes. Di gros sa, ren de ren sara tieu.
"Vai-t'en plus liuen rascla de bouto;
Crebo de fam l'iver, tu que cantes l'estieu."

Ansin charro la fablo antico
Per nous counseia la pratico
Di sarro-piastro, urous de nousa li cordoun
De si bourso.--Que la coulico

Rousigue la tripaio en aqueli coudoun!

Me fai susa, lou fabulisto,
Quand dis que l'iver vas en quisto
De mousco, verme, gran, tu que manges jamai.
De blad! Que n'en faries, ma fisto!
As ta fon melicouso e demandes ren mai.

Que t'enchau l'iver! Ta famiho
A la sousto en terro soumiho,
Et tu dormes la som que n'a ges de revei;
Toun cadabre toumbo en douliho.
Un jour, en tafurant, la fournigo lou vei,

De tu magro peu dessecado
La marriasso fai becado;
Te curo lou perus, te chapouto a mouceu,
T'encafourno per car-salado,
Requisto prouvisioun, l'iver, en tems de neu.

III.

Vaqui l'istori veritablo
Ben liuen dou conte de la fablo.
Que n'en pensas, caneu de sort!
--O rammaissaire de dardeno
Det croucu, boumbudo bedeno
Que gouvernas lou mounde eme lou coffre-fort,

Fases courre lou bru, canaio,
Que l'artisto jamai travaio
E deu pati, lou bedigas.
Teisas-vous dounc: quand di lambrusco

La Cigalo a cava la rusco,
Raubas soun beure, e piei, morto, la rousigas.

So speaks my friend in the expressive Provencal idiom, rehabilitating
the creature so libelled by the fabulist.

Translated with a little necessary freedom, the English of it is as
follows:--

I.

Fine weather for the Cigale! God, what heat!
  Half drunken with her joy, she feasts
In a hail of fire. Pays for the harvest meet;
  A golden sea the reaper breasts,
Loins bent, throat bare; silent, he labours long,
For thirst within his throat has stilled the song.

A blessed time for thee, little Cigale.
  Thy little cymbals shake and sound,
Shake, shake thy stomach till thy mirrors fall!
  Man meanwhile swings his scythe around;
Continually back and forth it veers,
Flashing its steel amidst the ruddy ears.

Grass-plugged, with water for the grinder full,
  A flask is hung upon his hip;
The stone within its wooden trough is cool,
  Free all the day to sip and sip;
But man is gasping in the fiery sun,
That makes his very marrow melt and run.

Thou, Cigale, hast a cure for thirst: the bark,
  Tender and juicy, of the bough.
Thy beak, a very needle, stabs it. Mark
  The narrow passage welling now;
The sugared stream is flowing, thee beside,
Who drinkest of the flood, the honeyed tide.

Not in peace always; nay, for thieves arrive,
  Neighbours and wives, or wanderers vile;
They saw thee sink the well, and ill they thrive
  Thirsting; they seek to drink awhile;
Beauty, beware! the wallet-snatcher's face,
Humble at first, grows insolent apace.

They seek the merest drop; thy leavings take;
  Soon discontent, their heads they toss;
They crave for all, and all will have. They rake
  Their claws thy folded wings across;
Thy back a mountain, up and down each goes;
They seize thee by the beak, the horns, the toes.

This way and that they pull. Impatient thou:
  Pst! Pst! a jet of nauseous taste
O'er the assembly sprinklest. Leave the bough
  And fly the rascals thus disgraced,
Who stole thy well, and with malicious pleasure
Now lick their honey'd lips, and feed at leisure.

See these Bohemians without labour fed!
  The ant the worst of all the crew--
Fly, drone, wasp, beetle too with horned head,
  All of them sharpers thro' and thro',
Idlers the sun drew to thy well apace--

None more than she was eager for thy place,

More apt thy face to tickle, toe to tread,
  Or nose to pinch, and then to run
Under the shade thine ample belly spread;
  Or climb thy leg for ladder; sun
Herself audacious on thy wings, and go
Most insolently o'er thee to and fro.

II.

Now comes a tale that no one should believe.
  In other times, the ancients say,
The winter came, and hunger made thee grieve.
  Thou didst in secret see one day
The ant below the ground her treasure store away.

The wealthy ant was drying in the sun
  Her corn the dew had wet by night,
Ere storing it again; and one by one
  She filled her sacks as it dried aright.
Thou camest then, and tears bedimmed thy sight,

Saying: "'Tis very cold; the bitter bise
  Blows me this way and that to-day.
I die of hunger. Of your riches please
  Fill me my bag, and I'll repay,
When summer and its melons come this way.

"Lend me a little corn." Go to, go to!
  Think you the ant will lend an ear?
You are deceived. Great sacks, but nought for you!

"Be off, and scrape some barrel clear!
You sing of summer: starve, for winter's here!"

'Tis thus the ancient fable sings
    To teach us all the prudence ripe
Of farthing-snatchers, glad to knot the string
    That tie their purses. May the gripe
Of colic twist the guts of all such tripe!

He angers me, this fable-teller does,
    Saying in winter thou dost seek
Flies, grubs, corn--thou dost never eat like us!
    --Corn! Couldst thou eat it, with thy beak?
Thou hast thy fountain with its honey'd reek.

To thee what matters winter? Underground
    Slumber thy children, sheltered; thou
The sleep that knows no waking sleepest sound.
    Thy body, fallen from the bough,
Crumbles; the questing ant has found thee now.

The wicked ant of thy poor withered hide
    A banquet makes; in little bits
She cuts thee up, and empties thine inside,
    And stores thee where in wealth she sits:
Choice diet when the winter numbs the wits.

III.

Here is the tale related duly,
And little resembling the fable, truly!
Hoarders of farthings, I know, deuce take it.

It isn't the story as you would make it!
Crook-fingers, big-bellies, what do you say,
Who govern the world with the cash-box--hey?

You have spread the story, with shrug and smirk,
That the artist ne'er does a stroke of work;
And so let him suffer, the imbecile!
Be you silent! 'Tis you, I think,
When the Cigale pierces the vine to drink,
Drive her away, her drink to steal;
And when she is dead--you make your meal!

CHAPTER II
THE CIGALE LEAVES ITS BURROW

The first Cigales appear about the summer solstice. Along the beaten paths, calcined by the sun, hardened by the passage of frequent feet, we see little circular orifices almost large enough to admit the thumb. These are the holes by which the larvae of the Cigale have come up from the depths to undergo metamorphosis. We see them more or less everywhere, except in fields where the soil has been disturbed by ploughing. Their usual position is in the driest and hottest situations, especially by the sides of roads or the borders of footpaths. Powerfully equipped for the purpose, able at need to pierce the turf or sun-dried clay, the larva, upon leaving the earth, seems to prefer the hardest spots.

A garden alley, converted into a little Arabia Petraea by reflection from a wall facing the south, abounds in such holes. During the last days of June I have made an examination of these recently abandoned pits. The soil is so compact that I needed a pick to tackle it.

The orifices are round, and close upon an inch in diameter. There is absolutely no debris round them; no earth thrown up from within. This is always the case; the holes of the Cigales are never surrounded by dumping-heaps, as are the burrows of the Geotrupes, another notable excavator. The way in which the work is done is responsible for this difference. The dung-beetle works from without inwards; she begins to dig at the mouth of the burrow, and afterwards re-ascends and accumulates the excavated material on the surface. The larva of the Cigale, on the contrary, works outward from within, upward from below; it opens the door of exit at the last moment, so that it is not free for the discharge of excavated material until the work is done. The first enters and raises a little rubbish-heap at the threshold of her burrow; the second emerges, and cannot, while working, pile up its rubbish on a threshold which as yet has no existence.

The burrow of the Cigale descends about fifteen inches. It is cylindrical, slightly twisted, according to the exigencies of the soil, and always approaches the vertical, or the direction of the shortest passage. It is perfectly free along its entire length. We shall search in vain for the rubbish which such an excavation must apparently produce; we shall find nothing of the sort. The burrow terminates in a cul-de-sac, in a fairly roomy chamber with unbroken walls, which shows not the least vestige of communication with any other burrow or prolongation of the shaft.

Taking its length and diameter into account, we find the excavation has a total volume of about twelve cubic inches. What becomes of the earth which is removed?

Sunk in a very dry, crumbling soil, we should expect the shaft and the chamber at the bottom to have soft, powdery walls, subject to petty landslips, if no work were done but that of excavation. On the contrary, the walls are neatly daubed, plastered with a sort of clay-like mortar. They are not precisely smooth, indeed they are distinctly rough; but their irregularities are covered with a layer of plaster, and the crumbling material, soaked in some glutinous liquid and dried, is held firmly in place.

The larva can climb up and down, ascend nearly to the surface, and go down into its chamber of refuge, without bringing down, with his claws, the continual falls of material which would block the burrow, make ascent a matter of difficulty, and retreat impossible. The miner shores up his galleries with uprights and cross-timbers; the builder of underground railways supports the sides and roofs of his tunnels with a lining of brick or masonry or segments of iron tube; the larva of the Cigale, no less prudent an engineer, plasters the walls of its burrow with cement, so that the passage is always free and ready for use.

If I surprise the creature just as it is emerging from the soil in order to gain a neighbouring bough and there undergo transformation, I see it immediately make a prudent retreat, descending to the bottom of its burrow without the slightest difficulty--a proof that even when about to be abandoned for ever the refuge is not encumbered with rubbish.

The ascending shaft is not a hurried piece of work, scamped by a creature impatient to reach the sunlight. It is a true dwelling, in which the larva may make a long stay. The plastered walls betray as much. Such precautions would be useless in the case of a simple exit abandoned as soon as made. We cannot doubt that the burrow is a kind of meteorological observatory, and that its inhabitant takes note of the weather without. Buried underground at a depth of twelve or fifteen inches, the larva, when ripe for escape, could hardly judge whether the meteorological conditions were favourable. The subterranean climate varies too little, changes too slowly, and would not afford it the precise information required for the most important action of its life--the escape into the sunshine at the time of metamorphosis.

Patiently, for weeks, perhaps for months, it digs, clears, and strengthens a vertical shaft, leaving only a layer of earth a finger's breadth in thickness to isolate it from the outer world. At the bottom it prepares a carefully built recess. This is its refuge, its place of waiting, where it reposes in peace if its observations decide it to postpone its final departure. At the least sign of fine weather it climbs to the top of its burrow, sounds the outer world through the thin layer of earth which covers the shaft, and informs itself of the temperature and humidity of the outer air.

If things are not going well--if there are threats of a flood or the dreaded *bise*--events of mortal gravity when the delicate insect issues from its cerements--the prudent creature re-descends to the bottom of its burrow for a longer wait. If, on the contrary, the state of the atmosphere is favourable, the roof is broken through by a few strokes of its claws, and the larva emerges from its tunnel.

Everything seems to prove that the burrow of the Cigale is a waiting-room, a meteorological station, in which the larva makes a prolonged stay; sometimes hoisting itself to the neighbourhood of the surface in order to ascertain the external climate; sometimes retiring to the depths the better to shelter itself. This explains the chamber at the base of the shaft, and the necessity of a cement to hold the walls together, for otherwise the creature's continual comings and goings would result in a landslip.

A matter less easy of explanation is the complete disappearance of the material which originally filled the excavated space. Where are the twelve cubic inches of earth that represent the average volume of the original contents of the shaft? There is not a trace of this material outside, nor inside either. And how, in a soil as dry as a cinder, is the plaster made with which the walls are covered?

Larvae which burrow in wood, such as those of Capricornis and Buprestes, will apparently answer our first question. They make their way through the substance of a tree-trunk, boring their galleries by the simple method of eating the material in front of them. Detached by their mandibles, fragment by fragment, the material is digested. It passes from end to end through the body of the pioneer, yields during its passage its meagre nutritive principles, and accumulates behind it, obstructing the passage, by which the larva will never return. The work of extreme division, effected partly by the mandibles and partly by the stomach, makes the digested material more compact than the intact wood, from which it follows that there is always a little free space at the head of the gallery, in which the caterpillar works and lives; it is not of any great length, but just suffices for the movements of the prisoner.

Must not the larva of the Cigale bore its passage in some such fashion? I do not mean that the results of excavation pass through its body--for earth, even the softest mould, could form no possible part of its diet. But is not the material detached simply thrust back behind the excavator as the work progresses?

The Cigale passes four years under ground. This long life is not spent, of course, at the bottom of the well I have just described; that is merely a resting-place preparatory to its appearance on the face of the earth. The larva comes from elsewhere; doubtless from a considerable distance. It is a vagabond, roaming from one root to another and implanting its rostrum. When it moves, either to flee from the upper layers of the soil, which in winter become too cold, or to install itself upon a more juicy root, it makes a road by rejecting behind it the material broken up by the teeth of its picks. That this is its method is incontestable.

As with the larvae of Capricornis and Buprestes, it is enough for the traveller to have around it the small amount of free space necessitated by its movements. Moist, soft, and easily compressible soil is to the larva of the Cigale what digested wood-pulp is to the others. It is compressed without difficulty, and so leaves a vacant space.

The difficulty is that sometimes the burrow of exit from the waiting-place is driven through a very arid soil, which is extremely refractory to compression so long as it retains its aridity. That the larva, when commencing the excavation of its burrow, has already thrust part of the detached material into a previously made gallery, now filled up and disappeared, is probable enough, although nothing in the actual condition of things goes to support the theory; but if we consider the capacity of the shaft and the extreme difficulty of making room for such a volume of debris, we feel dubious once more; for to hide such a quantity of earth a considerable empty space would be necessary, which could only be obtained by the disposal of more debris. Thus we are caught in a vicious circle. The mere packing of the powdered earth rejected behind the excavator would not account for so large a void. The Cigale must have a special method of disposing of the waste earth. Let us see if we can discover the secret.

Let us examine a larva at the moment of emerging from the soil. It is almost always more or less smeared with mud, sometimes dried, sometimes moist. The implements of excavation, the claws of the fore-feet, have their points covered by little globules of mortar; the others bear leggings of mud; the back is spotted with clay. One is reminded of a scavenger who has been scooping up mud all day. This condition is the more striking in that the insect comes from an absolutely dry soil. We should expect to see it dusty; we find it muddy.

One more step, and the problem of the well is solved. I exhume a larva which is working at its gallery of exit. Chance postpones this piece of luck, which I cannot expect to achieve at once, since nothing on the surface guides my search. But at last I am rewarded, and the larva is just beginning its excavation. An inch of tunnel, free of all waste or rubbish, and at the bottom the chamber, the place of rest; so far has the work proceeded. And the worker--in what condition is it? Let us see.

The larva is much paler in colour than those which I have caught as they emerged. The large eyes in particular are whitish, cloudy, blurred, and apparently blind. What would be the use of sight underground? The eyes of the larvae leaving their burrows are black and shining, and evidently capable of sight. When it issues into the sunlight the future Cigale must find, often at some distance from its burrow, a suitable twig from which to hang during its metamorphosis, so that sight is obviously of the greatest utility. The maturity of the eyes, attained during the time of preparation before deliverance, proves that the larva, far from boring its tunnel in haste, has spent a long time labouring at it.

What else do we notice? The blind, pale larva is far more voluminous than in the mature state; it is swollen with liquid as though it had dropsy. Taken in the fingers, a limpid serum oozes from the hinder part of the body, which moistens the whole surface. Is this fluid, evacuated by the intestine, a product of urinary secretion--simply the contents of a stomach nourished entirely upon sap? I will not attempt to decide, but for convenience will content myself with calling it urine.

Well, this fountain of urine is the key to the enigma. As it digs and advances the larva waters the powdery debris and converts it into a paste, which is immediately applied to the walls by the pressure of the abdomen. Aridity is followed by plasticity. The mud thus obtained penetrates the interstices of the rough soil; the more liquid portion enters the substance of the soil by infiltration; the remainder becomes tightly packed and fills up the inequalities of the walls. Thus the insect obtains an empty tunnel, with no loose waste, as all the loosened soil is utilised on the spot, converted into a mortar which is more compact and homogeneous than the soil through which the shaft is driven.

Thus the larva works in the midst of a coating of mud, which is the cause of its dirtiness, so astonishing when we see it issue from an excessively dry soil. The perfect insect, although henceforth liberated from the work of a sapper and miner, does not entirely abandon the use of urine as a weapon, employing it as a means of defence. Too closely observed it throws a jet of liquid upon the importunate enemy and flies away. In both its forms the Cigale, in spite of its dry temperament, is a famous irrigator.

Dropsical as it is, the larva cannot contain sufficient liquid to moisten and convert into easily compressible mud the long column of earth which must be removed from the burrow. The reservoir becomes exhausted, and the provision must be renewed. Where, and how? I think I can answer the question.

The few burrows uncovered along their entirety, with the meticulous care such a task demands, have revealed at the bottom, encrusted in the wall of the terminal chamber, a living root, sometimes of the thickness of a pencil, sometimes no bigger than a straw. The visible portion of this root is only a fraction of an inch in length; the rest is hidden by the surrounding earth. Is the presence of this source of sap fortuitous? Or is it the result of deliberate choice on the part of the larva? I incline towards the second alternative, so repeatedly was the presence of a root verified, at least when my search was skilfully conducted.

Yes, the Cigale, digging its chamber, the nucleus of the future shaft, seeks out the immediate neighbourhood of a small living root; it lays bare a certain portion, which forms part of the wall, without projecting. This living spot in the wall is the fountain where the supply of moisture is renewed. When its reservoir is exhausted by the conversion of dry dust into mud the miner descends to its chamber, thrusts its proboscis into the root, and drinks deep from the vat built into the wall. Its organs well filled, it re-ascends. It resumes work, damping the hard soil the better to remove it with its talons, reducing the debris to mud, in order to pack it tightly around it and obtain a free passage. In this manner the shaft is driven upwards; logic and the facts of the case, in the absence of direct observation, justify the assertion.

If the root were to fail, and the reservoir of the intestine were exhausted, what would happen? The following experiment will inform us: a larva is caught as it leaves the earth. I place it at the bottom of a test-tube, and cover it with a column of dry earth, which is rather lightly packed. This column is about six inches in height. The larva has just left an excavation three times as deep, made in soil of the same kind, but offering a far greater resistance. Buried under this short column of powdery earth, will it be able to gain the surface? If its strength hold out the issue should be certain; having but lately made its way through the hard earth, this obstacle should be easily removed.

But I am not so sure. In removing the stopper which divided it from the outside world, the larva has expended its final store of liquid. The cistern is dry, and in default of a living root there is no means of replenishing it. My suspicions are well founded. For three days the prisoner struggles desperately, but cannot ascend by so much as an inch. It is impossible to fix the material removed in the absence of moisture; as soon as it is thrust aside it slips back again. The labour has no visible result; it is a labour of Sisyphus, always to be commenced anew. On the fourth day the creature succumbs.

With the intestines full the result is very different.

I make the same experiment with an insect which is only beginning its work of liberation. It is swollen with fluid, which oozes from it and moistens the whole body. Its task is easy; the overlying earth offers little resistance. A small quantity of liquid from the intestines converts it into mud; forms a sticky paste which can be thrust aside with the assurance that it will remain where it is placed. The shaft is gradually opened; very unevenly, to be sure, and it is almost choked up behind the insect as it climbs upwards. It seems as though the creature recognises the impossibility of renewing its store of liquid, and so economises the little it possesses, using only just so much as is necessary in order to escape as quickly as possible from surroundings which are strange to its inherited instincts. This parsimony is so well judged that the insect gains the surface at the end of twelve days.

The gate of issue is opened and left gaping, like a hole made with an augur. For some little time the larva wanders about the neighbourhood of its burrow, seeking an eyrie on some low-growing bush or tuft of thyme, on a stem of grass or grain, or the twig of a shrub. Once found, it climbs and firmly clasps its support, the head upwards, while the talons of the fore feet close with an unyielding grip. The other claws, if the direction of the twig is convenient, assist in supporting it; otherwise the claws of the two fore legs will suffice. There follows a moment of repose, while the supporting limbs stiffen in an unbreakable hold. Then the thorax splits along the back, and through the fissure the insect slowly emerges. The whole process lasts perhaps half an hour.

There is the adult insect, freed of its mask, and how different from what it was but how! The wings are heavy, moist, transparent, with nervures of a tender green. The thorax is barely clouded with brown. All the rest of the body is a pale green, whitish in places. Heat and a prolonged air-bath are necessary to harden and colour the fragile creature. Some two hours pass without any perceptible change. Hanging to its deserted shell by the two fore limbs, the Cigale sways to the least breath of air, still feeble and still green. Finally, the brown colour appears and rapidly covers the whole body; the change of colour is completed in half an hour. Fastening upon its chosen twig at nine o'clock in the morning, the Cigale flies away under my eyes at half-past twelve.

The empty shell remains, intact except for the fissure in the back; clasping the twig so firmly that the winds of autumn do not always succeed in detaching it. For some months yet and even during the winter you will often find these forsaken skins hanging from the twigs in the precise attitude assumed by the larva at the moment of metamorphosis. They are of a horny texture, not unlike dry parchment, and do not readily decay.

I could gather some wonderful information regarding the Cigale were I to listen to all that my neighbours, the peasants, tell me. I will give one instance of rustic natural history.

Are you afflicted with any kidney trouble, or are you swollen with dropsy, or have you need of some powerful diuretic? The village pharmacopoeia is unanimous in recommending the Cigale as a sovereign remedy. The insects in the adult form are collected in summer. They are strung into necklaces which are dried in the sun and carefully preserved in some cupboard or drawer. A good housewife would consider it imprudent to allow July to pass without threading a few of these insects.

Do you suffer from any nephritic irritation or from stricture? Drink an infusion of Cigales. Nothing, they say, is more effectual. I must take this opportunity of thanking the good soul who once upon a time, so I was afterwards informed, made me drink such a concoction unawares for the cure of some such trouble; but I still remain incredulous. I have been greatly struck by the fact that the ancient physician of Anazarbus used to recommend the same remedy. Dioscorides tells us: *Cicadae, quae inassatae manduntur, vesicae doloribus prosunt*. Since the distant days of this patriarch of *materia medica* the Provencal peasant has retained his faith in the remedy revealed to him by the Greeks, who came from Phocaea with the olive, the fig, and the vine. Only one thing is changed: Dioscorides advises us to eat the Cigales roasted, but now they are boiled, and the decoction is administered as medicine. The explanation which is given of the diuretic properties of the insect is a marvel of ingenuousness. The Cigale, as every one knows who has tried to catch it, throws a jet of liquid excrement in one's face as it flies away. It therefore endows us with its faculties of evacuation. Thus Dioscorides and his contemporaries must have reasoned; so reasons the peasant of Provence to-day.

What would you say, worthy neighbours, if you knew of the virtues of the larva, which is able to mix sufficient mortar with its urine to build a meteorological station and a shaft connecting with the outer world? Your powers should equal those of Rabelais' Gargantua, who, seated upon the towers of Notre Dame, drowned so many thousands of the inquisitive Parisians.

CHAPTER III
THE SONG OF THE CIGALE

Where I live I can capture five species of Cigale, the two principal species being the common Cigale and the variety which lives on the flowering ash. Both of these are widely distributed and are the only species known to the country folk. The larger of the two is the common Cigale. Let me briefly describe the mechanism with which it produces its familiar note.

On the under side of the body of the male, immediately behind the posterior limbs, are two wide semicircular plates which slightly overlap one another, the right hand lying over the left hand plate. These are the shutters, the lids, the dampers of the musical-box. Let us remove them. To the right and left lie two spacious cavities which are known in Provencal as the chapels (li capello). Together they form the church (la gleiso). Their forward limit is formed by a creamy yellow membrane, soft and thin; the hinder limit by a dry membrane coloured like a soap bubble and known in Provencal as the mirror (mirau).

The church, the mirrors, and the dampers are commonly regarded as the organs which produce the cry of the Cigale. Of a singer out of breath one says that he has broken his mirrors (a li mirau creba). The same phrase is used of a poet without inspiration. Acoustics give the lie to the popular belief. You may break the mirrors, remove the covers with a snip of the scissors, and tear the yellow anterior membrane, but these mutilations do not silence the song of the Cigale; they merely change its quality and weaken it. The chapels are resonators; they do not produce the sound, but merely reinforce it by the vibration of their anterior and posterior membranes; while the sound is modified by the dampers as they are opened more or less widely.

The actual source of the sound is elsewhere, and is somewhat difficult for a novice to find. On the outer wall of either chapel, at the ridge formed by the junction of back and belly, is a tiny aperture with a horny circumference masked by the overlapping damper. We will call this the window. This opening gives access to a cavity or sound-chamber, deeper than the "chapels," but of much smaller capacity. Immediately behind the attachment of the posterior wings is a slight protuberance, almost egg-shaped, which is distinguishable, on account of its dull black colour, from the neighbouring integuments, which are covered with a silvery down. This protuberance is the outer wall of the sound-chamber.

Let us cut it boldly away. We shall then lay bare the mechanism which produces the sound, the *cymbal*. This is a small dry, white membrane, oval in shape, convex on the outer side, and crossed along its larger diameter by a bundle of three or four brown nervures, which give it elasticity. Its entire circumference is rigidly fixed. Let us suppose that this convex scale is pulled out of shape from the interior, so that it is slightly flattened and as quickly released; it will immediately regain its original convexity owing to the elasticity of the nervures. From this oscillation a ticking sound will result.

Twenty years ago all Paris was buying a silly toy, called, I think, the cricket or *cri-cri*. It was a short slip of steel fixed by one end to a metallic base. Pressed out of shape by the thumb and released, it yielded a very distressing, tinkling *click*. Nothing else was needed to take the popular mind by storm. The "cricket" had its day of glory. Oblivion has executed justice upon it so effectually that I fear I shall not be understood when I recall this celebrated device.

The membranous cymbal and the steel cricket are analogous instruments. Both produce a sound by reason of the rapid deformation and recovery of an elastic substance--in one case a convex membrane; in the other a slip of steel. The "cricket" was bent out of shape by the thumb. How is the convexity of the cymbals altered? Let us return to the "church" and break down the yellow curtain which closes the front of each chapel. Two thick muscular pillars are visible, of a pale orange colour; they join at an angle, forming a ˜V˜, of which the point lies on the median line of the insect, against the lower face of the thorax. Each of these pillars of flesh terminates suddenly at its upper extremity, as though cut short, and from the truncated portion rises a short, slender tendon, which is attached laterally to the corresponding cymbal.

There is the whole mechanism, no less simple than that of the steel "cricket." The two muscular columns contract and relax, shorten and lengthen. By means of its terminal thread each sounds its cymbal, by depressing it and immediately releasing it, when its own elasticity makes it spring back into shape. These two vibrating scales are the source of the Cigale's cry.

Do you wish to convince yourself of the efficiency of this mechanism? Take a Cigale but newly dead and make it sing. Nothing is simpler. Seize one of these muscular columns with the forceps and pull it in a series of careful jerks. The extinct *cri-cri* comes to life again; at each jerk there is a clash of the cymbal. The sound is feeble, to be sure, deprived of the amplitude which the living performer is able to give it by means of his resonating chambers; none the less, the fundamental element of the song is produced by this anatomist's trick.

Would you, on the other hand, silence a living Cigale?--that obstinate melomaniac, who, seized in the fingers, deplores his misfortune as loquaciously as ever he sang the joys of freedom in his tree? It is useless to violate his chapels, to break his mirrors; the atrocious mutilation would not quiet him. But introduce a needle by the lateral aperture which we have named the "window" and prick the cymbal at the bottom of the sound-box. A little touch and the perforated cymbal is silent. A similar operation on the other side of the insect and the insect is dumb, though otherwise as vigorous as before and without any perceptible wound. Any one not in the secret would be amazed at the result of my pin-prick, when the destruction of the mirrors and the other dependencies of the "church" do not cause silence. A tiny perforation of no importance to the insect is more effectual than evisceration.

The dampers, which are rigid and solidly built, are motionless. It is the abdomen itself which, by rising and falling, opens or closes the doors of the "church." When the abdomen is lowered the dampers exactly cover the chapels as well as the windows of the sound-boxes. The sound is then muted, muffled, diminished. When the abdomen rises the chapels are open, the windows unobstructed, and the sound acquires its full volume. The rapid oscillations of the abdomen, synchronising with the contractions of the motor muscles of the cymbals, determine the changing volume of the sound, which seems to be caused by rapidly repeated strokes of a fiddlestick.

If the weather is calm and hot, towards mid-day the song of the Cigale is divided into strophes of several seconds' duration, which are separated by brief intervals of silence. The strophe begins suddenly. In a rapid crescendo, the abdomen oscillating with increasing rapidity, it acquires its maximum volume; it remains for a few seconds at the same degree of intensity, then becomes weaker by degrees, and degenerates into a shake, which decreases as the abdomen returns to rest. With the last pulsations of the belly comes silence; the length of the silent interval varies according to the state of the atmosphere. Then, of a sudden, begins a new strophe, a monotonous repetition of the first; and so on indefinitely.

It often happens, especially during the hours of the sultry afternoons, that the insect, intoxicated with sunlight, shortens and even suppresses the intervals of silence. The song is then continuous, but always with an alternation of crescendo and diminuendo. The first notes are heard about seven or eight o'clock in the morning, and the orchestra ceases only when the twilight fails, about eight o'clock at night. The concert lasts a whole round of the clock. But if the sky is grey and the wind chilly the Cigale is silent.

The second species, only half the size of the common Cigale, is known in Provence as the **Cacan**; the name, being a fairly exact imitation of the sound emitted by the insect. This is the Cigale of the flowering ash, far more alert and far more suspicious than the common species. Its harsh, loud song consists of a series of cries--can! can! can! can!--with no intervals of silence subdividing the poem into stanzas. Thanks to its monotony and its harsh shrillness, it is a most odious sound, especially when the orchestra consists of hundreds of performers, as is often the case in my two plane-trees during the dog-days. It is as though a heap of dry walnuts were being shaken up in a bag until the shells broke. This painful concert, which is a real torment, offers only one compensation: the Cigale of the flowering ash does not begin his song so early as the common Cigale, and does not sing so late in the evening.

Although constructed on the same fundamental principles, the vocal organs exhibit a number of peculiarities which give the song its special character. The sound-box is lacking, which suppresses the entrance to it, or the window. The cymbal is uncovered, and is visible just behind the attachment of the hinder wing. It is, as before, a dry white scale, convex on the outside, and crossed by a bundle of fine reddish-brown nervures.

From the forward side of the first segment of the abdomen project two short, wide, tongue-shaped projections, the free extremities of which rest on the cymbals. These tongues may be compared to the blade of a watchman's rattle, only instead of engaging with the teeth of a rotating wheel they touch the nervures of the vibrating cymbal. From this fact, I imagine, results the harsh, grating quality of the cry. It is hardly possible to verify the fact by holding the insect in the fingers; the terrified *Cacan* does not go on singing his usual song.

The dampers do not overlap; on the contrary, they are separated by a fairly wide interval. With the rigid tongues, appendages of the abdomen, they half shelter the cymbals, half of which is completely bare. Under the pressure of the finger the abdomen opens a little at its articulation with the thorax. But the insect is motionless when it sings; there is nothing of the rapid vibrations of the belly which modulate the song of the common Cigale. The chapels are very small; almost negligible as resonators. There are mirrors, as in the common Cigale, but they are very small; scarcely a twenty-fifth of an inch in diameter. In short, the resonating mechanism, so highly developed in the common Cigale, is here extremely rudimentary. How then is the feeble vibration of the cymbals re enforced until it becomes intolerable?

This species of Cigale is a ventriloquist. If we examine the abdomen by transmitted light, we shall see that the anterior two-thirds of the abdomen are translucent. With a snip of the scissors we will cut off the posterior third, to which are relegated, reduced to the strictly indispensable, the organs necessary to the propagation of the species and the preservation of the individual. The rest of the abdomen presents a spacious cavity, and consists simply of the integuments of the walls, except on the dorsal side, which is lined with a thin muscular layer, and supports a fine digestive canal, almost a thread. This large cavity, equal to nearly half the total volume of the insect, is thus almost absolutely empty. At the back are seen the two motor muscles of the cymbals, two muscular columns arranged like the limbs of a ˜V˜. To right and left of the point of this ˜V˜ shine the tiny mirrors; and between the two branches of muscle the empty cavity is prolonged into the depths of the thorax.

This empty abdomen with its thoracic annex forms an enormous resonator, such as no other performer in our countryside can boast of. If I close with my finger the orifice of the truncated abdomen the sound becomes flatter, in conformity with the laws affecting musical resonators; if I fit into the aperture of the open body a tube or trumpet of paper the sound grows louder as well as deeper. With a paper cone corresponding to the pitch of the note, with its large end held in the mouth of a test-tube acting as a resonator, we have no longer the cry of the Cigale, but almost the bellowing of a bull. My little children, coming up to me by chance at the moment of this acoustic experiment, fled in terror.

The grating quality of the sound appears to be due to the little tongues which press on the nervures of the vibrating cymbals; the cause of its intensity is of course the ample resonator in the abdomen. We must admit that one must truly have a real passion for song before one would empty one's chest and stomach in order to make room for a musical-box. The necessary vital organs are extremely small, confined to a mere corner of the body, in order to increase the amplitude of the resonating cavity. Song comes first of all; other matters take the second rank.

It is lucky that the **Cacan** does not follow the laws of evolution. If, more enthusiastic in each generation, it could acquire, in the course of progress, a ventral resonator comparable to my paper trumpets, the South of France would sooner or later become uninhabitable, and the **Cacan** would have Provence to itself.

After the details already given concerning the common Cigale it is hardly needful to tell you how the insupportable **Cacan** can be reduced to silence. The cymbals are plainly visible on the exterior. Pierce them with the point of a needle, and immediately you have perfect silence. If only there were, in my plane-trees, among the insects which carry gimlets, some friends of silence like myself, who would devote themselves to such a task! But no: a note would be lacking in the majestic symphony of harvest-tide.

We are now familiar with the structure of the musical organ of the Cigale. Now the question arises: What is the object of these musical orgies? The reply seems obvious: they are the call of the males inviting their mates; they constitute a lovers' cantata.

I am going to consider this reply, which is certainly a very natural one. For thirty years the common Cigale and his unmusical friend the **Cacan** have thrust their society upon me. For two months every summer I have them under my eyes, and their voice in my ears. If I do not listen to them very willingly I observe them with considerable zeal. I see them ranged in rows on the smooth rind of the plane-trees, all with their heads uppermost, the two sexes mingled, and only a few inches apart.

The proboscis thrust into the bark, they drink, motionless. As the sun moves, and with it the shadow, they also move round the branch with slow lateral steps, so as to keep upon that side which is most brilliantly illuminated, most fiercely heated. Whether the proboscis is at work or not the song is never interrupted.

Now are we to take their interminable chant for a passionate love-song? I hesitate. In this gathering the two sexes are side by side. One does not spend months in calling a person who is at one's elbow. Moreover, I have never seen a female rush into the midst of even the most deafening orchestra. Sight is a sufficient prelude to marriage, for their sight is excellent. There is no need for the lover to make an everlasting declaration, for his mistress is his next-door neighbour.

Is the song a means of charming, of touching the hard of heart? I doubt it. I observe no sign of satisfaction in the females; I have never seen them tremble or sway upon their feet, though their lovers have clashed their cymbals with the most deafening vigour.

My neighbours the peasants say that at harvest-time the Cigale sings to them: *Sego, sego, sego!* (Reap, reap, reap!) to encourage them in their work. Harvesters of ideas and of ears of grain, we follow the same calling; the latter produce food for the stomach, the former food for the mind. Thus I understand their explanation and welcome it as an example of gracious simplicity.

Science asks for a better explanation, but finds in the insect a world which is closed to us. There is no possibility of foreseeing, or even of suggesting the impression produced by this clashing of cymbals upon those who inspire it. The most I can say is that their impassive exterior seems to denote a complete indifference. I do not insist that this is so; the intimate feelings of the insect are an insoluble mystery.

Another reason for doubt is this: all creatures affected by song have acute hearing, and this sense of hearing, a vigilant sentinel, should give warning of danger at the slightest sound. The birds have an exquisite delicacy of hearing. If a leaf stirs among the branches, if two passers-by exchange a word, they are suddenly silent, anxious, and on their guard. But the Cigale is far from sharing in such emotions. It has excellent sight. Its great faceted eyes inform it of all that happens to right and left; its three stemmata, like little ruby telescopes, explore the sky above its head. If it sees us coming it is silent at once, and flies away. But let us get behind the branch on which it is singing; let us manoeuvre so as to avoid the five centres of vision, and then let us speak, whistle, clap the hands, beat two stones together. For far less a bird which could not see you would stop its song and fly away terrified. The Cigale imperturbably continues to sing as if nothing had occurred.

Of my experiences of this kind I will mention only one, the most remarkable of many.

I borrowed the municipal artillery; that is, the iron boxes which are charged with gunpowder on the day of the patron saint. The artilleryman was delighted to load them for the benefit of the Cigales, and to fire them off for me before my house. There were two of these boxes stuffed full of powder as though for the most solemn rejoicing. Never was politician making his electoral progress favoured with a bigger charge. To prevent damage to my windows the sashes were all left open. The two engines of detonation were placed at the foot of the plane-trees before my door, no precautions being taken to mask them. The Cigales singing in the branches above could not see what was happening below.

There were six of us, spectators and auditors. We waited for a moment of relative quiet. The number of singers was counted by each of us, as well as the volume and rhythm of the song. We stood ready, our ears attentive to the aerial orchestra. The box exploded with a clap of thunder.

No disturbance ensued above. The number of performers was the same, the rhythm the same, the volume the same. The six witnesses were unanimous: the loud explosion had not modified the song of the Cigales in the least. The second box gave an identical result.

What are we to conclude from this persistence of the orchestra, its lack of surprise or alarm at the firing of a charge? Shall we conclude that the Cigale is deaf? I am not going to venture so far as that; but if any one bolder than myself were to make the assertion I really do not know what reasons I could invoke to disprove it. I should at least be forced to admit that it is very hard of hearing, and that we may well apply to it the homely and familiar phrase: to shout like a deaf man.

When the blue-winged cricket, basking on the pebbles of some country footpath, grows deliciously intoxicated with the heat of the sun and rubs its great posterior thighs against the roughened edge of its wing-covers; when the green tree-frog swells its throat in the foliage of the bushes, distending it to form a resonant cavity when the rain is imminent, is it calling to its absent mate? By no means. The efforts of the former produce a scarcely perceptible stridulation; the palpitating throat of the latter is as ineffectual; and the desired one does not come.

Does the insect really require to emit these resounding effusions, these vociferous avowals, in order to declare its passion? Consult the immense majority whom the conjunction of the sexes leaves silent. In the violin of the grasshopper, the bagpipe of the tree-frog, and the cymbals of the **Cacan** I see only their peculiar means of expressing the joy of living, the universal joy which every species of animal expresses after its kind.

If you were to tell me that the Cigales play on their noisy instruments careless of the sound produced, and merely for the pleasure of feeling themselves alive, just as we rub our hands in a moment of satisfaction, I should not be particularly shocked. That there is a secondary object in their conceit, in which the silent sex is interested, is very possible and very natural, but it is not as yet proven.[1]

CHAPTER IV
THE CIGALE. THE EGGS AND THEIR HATCHING

The Cigale confides its eggs to dry, slender twigs. All the branches examined by Reaumur which bore such eggs were branches of the mulberry: a proof that the person entrusted with the search for these eggs in the neighbourhood of Avignon did not bring much variety to his quest. I find these eggs not only on the mulberry-tree, but on the peach, the cherry, the willow, the Japanese privet, and other trees. But these are exceptions; what the Cigale really prefers is a slender twig of a thickness varying from that of a straw to that of a pencil. It should have a thin woody layer and plenty of pith. If these conditions are fulfilled the species matters little. I should pass in review all the semi-ligneous plants of the country were I to catalogue the various supports which are utilised by the gravid female.

Its chosen twig never lies along the ground; it is always in a more or less vertical position. It is usually growing in its natural position, but is sometimes detached; in the latter case it will by chance have fallen so that it retains its upright position. The insect prefers a long, smooth, regular twig which can receive the whole of its eggs. The best batches of eggs which I have found have been laid upon twigs of the *Spartium junceum*, which are like straws stuffed with pith, and especially on the upper twigs of the *Asphodelus cerasiferus*, which rises nearly a yard from the ground before ramifying.

It is essential that the support, no matter what its nature, should be dead and perfectly dry.

The first operation performed by the Cigale consists in making a series of slight lacerations, such as one might make with the point of a pin, which, if plunged obliquely downwards into the twig, would tear the woody fibres and would compress them so as to form a slight protuberance.

If the twig is irregular in shape, or if several Cigales have been working successively at the same point, the distribution of the punctures is confused; the eye wanders, incapable of recognising the order of their succession or the work of the individual. One characteristic is always present, namely, the oblique direction of the woody fragment which is raised by the perforation, showing that the Cigale always works in an upright position and plunges its rostrum downwards in the direction of the twig.

If the twig is regular, smooth, and conveniently long the perforations are almost equidistant and lie very nearly in a straight line. Their number varies; it is small when the mother, disturbed in her operations, has flown away to continue her work elsewhere; but they number thirty or forty, more or less, when they contain the whole of her eggs.

Each one of the perforations is the entrance to an oblique tunnel, which is bored in the medullary sheath of the twig. The aperture is not closed, except by the bunch of woody fibres, which, parted at the moment when the eggs are laid, recover themselves when the double saw of the oviduct is removed. Sometimes, but by no means always, you may see between the fibres a tiny glistening patch like a touch of dried white of egg. This is only an insignificant trace of some albuminous secretion accompanying the egg or facilitating the work of the double saw of the oviduct.

Immediately below the aperture of the perforation is the egg chamber: a short, tunnel-shaped cavity which occupies almost the whole distance between one opening and that lying below it. Sometimes the separating partition is lacking, and the various chambers run into one another, so that the eggs, although introduced by the various apertures, are arranged in an uninterrupted row. This arrangement, however, is not the most usual.

The contents of the chambers vary greatly. I find in each from six to fifteen eggs. The average is ten. The total number of chambers varying from thirty to forty, it follows that the Cigale lays from three to four hundred eggs. Reaumur arrived at the same figures from an examination of the ovaries.

This is truly a fine family, capable by sheer force of numbers of surviving the most serious dangers. I do not see that the adult Cigale is exposed to greater dangers than any other insect: its eye is vigilant, its departure sudden, and its flight rapid; and it inhabits heights at which the prowling brigands of the turf are not to be feared. The sparrow, it is true, will greedily devour it. From time to time he will deliberately and meditatively descend upon the plane-trees from the neighbouring roof and snatch up the singer, who squeaks despairingly. A few blows of the beak and the Cigale is cut into quarters, delicious morsels for the nestlings. But how often does the bird return without his prey! The Cigale, foreseeing his attack, empties its intestine in the eyes of its assailant and flies away.

But the Cigale has a far more terrible enemy than the sparrow. This is the green grasshopper. It is late, and the Cigales are silent. Drowsy with light and heat, they have exhausted themselves in producing their symphonies all day long. Night has come, and with it repose; but a repose frequently troubled. In the thick foliage of the plane-trees there is a sudden sound like a cry of anguish, short and strident. It is the despairing lamentation of the Cigale surprised in the silence by the grasshopper, that ardent hunter of the night, which leaps upon the Cigale, seizes it by the flank, tears it open, and devours the contents of the stomach. After the orgy of music comes night and assassination.

I obtained an insight into this tragedy in the following manner: I was walking up and down before my door at daybreak when something fell from the neighbouring plane-tree uttering shrill squeaks. I ran to see what it was. I found a green grasshopper eviscerating a struggling Cigale. In vain did the latter squeak and gesticulate; the other never loosed its hold, but plunged its head into the entrails of the victim and removed them by little mouthfuls.

This was instructive. The attack was delivered high up above my head, in the early morning, while the Cigale was resting; and the struggles of the unfortunate creature as it was dissected alive had resulted in the fall of assailant and assailed together. Since then I have often been the witness of similar assassinations.

I have even seen the grasshopper, full of audacity, launch itself in pursuit of the Cigale, who fled in terror. So the sparrow-hawk pursues the skylark in the open sky. But the bird of prey is less ferocious than the insect; it pursues a creature smaller than itself. The locust, on the contrary, assails a colossus, far larger and far more vigorous than its enemy; yet the result is a foregone conclusion, in spite of this disproportion. With its powerful mandibles, like pincers of steel, the grasshopper rarely fails to eviscerate its captive, which, being weaponless, can only shriek and struggle.

The Cigale is an easy prey during its hours of somnolence. Every Cigale encountered by the ferocious grasshopper on its nocturnal round must miserably perish. Thus are explained those sudden squeaks of anguish which are sometimes heard in the boughs during the hours of the night and early morning, although the cymbals have long been silent. The sea-green bandit has fallen upon some slumbering Cigale. When I wished to rear some green grasshoppers I had not far to seek for the diet of my pensioners; I fed them on Cigales, of which enormous numbers were consumed in my breeding-cages. It is therefore an established fact that the green grasshopper, the false Cigale of the North, will eagerly devour the true Cigale, the inhabitant of the Midi.

But it is neither the sparrow nor the green grasshopper that has forced the Cigale to produce such a vast number of offspring. The real danger is elsewhere, as we shall see. The risk is enormous at the moment of hatching and also when the egg is laid.

Two or three weeks after its escape from the earth--that is, about the middle of July--the Cigale begins to lay. In order to observe the process without trusting too much to chance, I took certain precautions which would, I felt sure, prove successful. The dry Asphodelus is the support preferred by the insect, as previous observations had assured me. It was also the plant which best lent itself to my experiments, on account of its long, smooth stems. Now, during the first years of my residence in the South I replaced the thistles in my paddock by other native plants of a less stubborn and prickly species. Among the new occupants was the asphodel. This was precisely what I needed for my experiments. I left the dry stems of the preceding year in place, and when the breeding season arrived I inspected them daily.

I had not long to wait. As early as July 15th I found as many Cigales as I could wish on the stems of the asphodel, all in process of laying. The gravid female is always solitary. Each mother has her twig to herself, and is in no danger of being disturbed during the delicate operation of laying. When the first occupant has departed another may take her place, and so on indefinitely. There is abundance of room for all; but each prefers to be alone as her turn arrives. There is, however, no unpleasantness of any kind; everything passes most peacefully. If a female Cigale finds a place which has been already taken she flies away and seeks another twig directly she discovers her mistake.

The gravid female always retains an upright position at this time, as indeed she does at other times. She is so absorbed in her task that she may readily be watched, even through a magnifying glass. The ovipositor, which is about four-tenths of an inch in length, is plunged obliquely and up to the hilt into the twig. So perfect is the tool that the operation is by no means troublesome. We see the Cigale tremble slightly, dilating and contracting the extremity of the abdomen in frequent palpitations. This is all that can be seen. The boring instrument, consisting of a double saw, alternately rises and sinks in the rind of the twig with a gentle, almost imperceptible movement. Nothing in particular occurs during the process of laying the eggs. The insect is motionless, and hardly ten minutes elapse between the first cut of the ovipositor and the filling of the egg-chamber with eggs.

The ovipositor is then withdrawn with methodical deliberation, in order that it may not be strained or bent. The egg-chamber closes of its own accord as the woody fibres which have been displaced return to their position, and the Cigale climbs a little higher, moving upwards in a straight line, by about the length of its ovipositor. It then makes another puncture and a fresh chamber for another ten or twelve eggs. In this way it scales the twig from bottom to top.

These facts being understood, we are able to explain the remarkable arrangement of the eggs. The openings in the rind of the twig are practically equidistant, since each time the Cigale moves upward it is by a given length, namely, that of the ovipositor. Very rapid in flight, she is a very idle walker. At the most you may see her, on the living twig from which she is drinking, moving at a slow, almost solemn pace, to gain a more sunny point close at hand. On the dry twig in which she deposits her eggs she observes the same formal habits, and even exaggerates them, in view of the importance of the operation. She moves as little as possible, just so far as she must in order to avoid running two adjacent egg-chambers into one. The extent of each movement upwards is approximately determined by the depth of the perforation.

The apertures are arranged in a straight line when their number is not very large. Why, indeed, should the insect wander to right or to left upon a twig which presents the same surface all over? A lover of the sun, she chooses that side of the twig which is most exposed to it. So long as she feels the heat, her supreme joy, upon her back, she will take good care not to change the position which she finds so delightful for another in which the sun would fall upon her less directly.

The process of depositing the eggs is a lengthy one when it is carried out entirely on the same twig. Counting ten minutes for each egg-chamber, the full series of forty would represent a period of six or seven hours. The sun will of course move through a considerable distance before the Cigale can finish her work. In such cases the series of apertures follows a spiral curve. The insect turns round the stalk as the sun turns.

Very often as the Cigale is absorbed in her maternal task a diminutive fly, also full of eggs, busily exterminates the Cigale's eggs as fast as they are laid.

This insect was known to Reaumur. In nearly all the twigs examined he found its grub, the cause of a misunderstanding at the beginning of his researches. But he did not, could not see the audacious insect at work. It is one of the Chalcididae, about one-fifth or one-sixth of an inch in length; entirely black, with knotty antennae, which are slightly thicker towards their extremities. The unsheathed ovipositor is implanted in the under portion of the abdomen, about the middle, and at right angles to the axis of the body, as in the case of the Leucospis, the pest of the apiary. Not having taken the precaution to capture it, I do not know what name the entomologists have bestowed upon it, or even if this dwarf exterminator of the Cigale has as yet been catalogued. What I am familiar with is its calm temerity, its impudent audacity in the presence of the colossus who could crush it with a foot. I have seen as many as three at once exploiting the unfortunate female. They keep close behind the Cigale, working busily with their probes, or waiting until their victim deposits her eggs.

The Cigale fills one of her egg-chambers and climbs a little higher in order to bore another hole. One of the bandits runs to the abandoned station, and there, almost under the claws of the giant, and without the least nervousness, as if it were accomplishing some meritorious action, it unsheathes its probe and thrusts it into the column of eggs, not by the open aperture, which is bristling with broken fibres, but by a lateral fissure. The probes works slowly, as the wood is almost intact. The Cigale has time to fill the adjacent chamber.

As soon as she has finished one of these midges, the very same that has been performing its task below her, replaces her and introduces its disastrous egg. By the time the Cigale departs, her ovaries empty, the majority of the egg-chambers have thus received the alien egg which will work the destruction of their contents. A small, quick-hatching grub, richly nourished on a dozen eggs, will replace the family of the Cigale.

The experience of centuries has taught the Cigale nothing. With her excellent eyesight she must be able to perceive these terrible sappers as they hover about her, meditating their crime. Too peaceable giantess! if you see them why do you not seize them in your talons, crush the pigmies at their work, so that you may proceed with your travail in security? But no, you will leave them untouched; you cannot modify your instincts, even to alleviate your maternal misfortunes.

The eggs of the common Cigale are of a shining ivory white. Conical at the ends, and elongated in form, they might be compared in shape to the weaver's shuttle. Their length is about one-tenth of an inch, their diameter about one-fiftieth. They are packed in a row, slightly overlapping one another. The eggs of the Cacan are slightly smaller, and are assembled in regular groups which remind one of microscopical bundles of cigars. We will consider the eggs of the common Cigale to the exclusion of the others, as their history is the history of all.

September is not yet over when the shining white as of ivory gives way to the yellow hue of cheese. During the first days of October you may see, at the forward end of the egg, two tiny points of chestnut brown, which are the eyes of the embryo in formation. These two shining eyes, which almost seem to gaze at one, and the cone-shaped head of the egg, give it the look of a tiny fish without fins--a fish for whom half a nut-shell would make a capacious aquarium.

About the same time I notice frequently, on the asphodels in the paddock and on those of the neighbouring hills, certain indications that the eggs have recently hatched out. There are certain cast-off articles of clothing, certain rags and tatters, left on the threshold of the egg-chamber by the new-born grubs as they leave it and hurry in search of a new lodging. We shall see in a moment what these vestiges mean.

But in spite of my visits, which were so assiduous as to deserve success, I had never contrived to see the young Cigales emerge from their egg-chambers. My domestic researches had been pursued in vain. Two years running I had collected, in boxes, tubes, and bottles, a hundred twigs of every kind which were peopled by the eggs of the Cigale; but not one had shown me what I so desired to witness: the issue of the new-born Cigales.

Reaumur experienced the same disappointment. He tells us how all the eggs supplied by his friends were abortive, even when he placed them in a glass tube thrust under his armpit, in order to keep them at a high temperature. No, venerable master! neither the temperate shelter of our studies and laboratories, nor the incubating warmth of our bodies is sufficient here; we need the supreme stimulant, the kiss of the sun; after the cool of the mornings, which are already sharp, the sudden blaze of the superb autumn weather, the last endearments of summer.

It was under such circumstances, when a blazing sun followed a cold night, that I found the signs of completed incubation; but I always came too late; the young Cigales had departed. At most I sometimes found one hanging by a thread to its natal stem and struggling in the air. I supposed it to be caught in a thread of gossamer, or some shred of cobweb.

At last, on the 27th of October, despairing of success, I gathered some asphodels from the orchard, and the armful of dry twigs in which the Cigales had laid their eggs was taken up to my study. Before giving up all hope I proposed once more to examine the egg-chambers and their contents. The morning was cold, and the first fire of the season had been lit in my room. I placed my little bundle on a chair before the fire, but without any intention of testing the effect of the heat of the flames upon the concealed eggs. The twigs, which I was about to cut open, one by one, were placed there to be within easy reach of my hand, and for no other reason.

Then, while I was examining a split twig with my magnifying-glass, the phenomenon which I had given up all hope of observing took place under my eyes. My bundle of twigs was suddenly alive; scores and scores of the young larvae were emerging from their egg-chambers. Their numbers were such that my ambition as observer was amply satisfied. The eggs were ripe, on the point of hatching, and the warmth of the fire, bright and penetrating, had the effect of sunlight in the open. I was quick to profit by the unexpected piece of good fortune.

At the orifice of the egg-chamber, among the torn fibres of the bark, a little cone-shaped body is visible, with two black eye-spots; in appearance it is precisely like the fore portion of the butter-coloured egg; or, as I have said, like the fore portion of a tiny fish. You would think that an egg had been somehow displaced, had been removed from the bottom of the chamber to its aperture. An egg to move in this narrow passage! a walking egg! No, that is impossible; eggs "do not do such things!" This is some mistake. We will break open the twig, and the mystery is unveiled. The actual eggs are where they always were, though they are slightly disarranged. They are empty, reduced to the condition of transparent skins, split wide open at the upper end. From them has issued the singular organism whose most notable characteristics are as follows:--

In its general form, the configuration of the head and the great black eyes, the creature, still more than the egg, has the appearance of an extremely minute fish. A simulacrum of a ventral fin increases the resemblance. This apparent fin in reality consists of the two fore-limbs, which, packed in a special sheath, are bent backwards, stretched out against one another in a straight line. Its small degree of mobility must enable the grub to escape from the egg-shell and, with greater difficulty, from the woody tunnel leading to the open air. Moving outwards a little from the body, and then moving back again, this lever serves as a means of progression, its terminal hooks being already fairly strong. The four other feet are still covered by the common envelope, and are absolutely inert. It is the same with the antennae, which can scarcely be seen through the magnifying-glass. The organism which has issued from the egg is a boat-shaped body with a fin-shaped limb pointing backwards on the ventral face, formed by the junction of the two fore-limbs. The segmentation of the body is very clear, especially on the abdomen. The whole body is perfectly smooth, without the least suspicion of hair.

What name are we to give to this initial phase of the Cigale--a phase so strange, so unforeseen, and hitherto unsuspected? Must I amalgamate some more or less appropriate words of Greek and fabricate a portentous nomenclature? No, for I feel sure that barbarous alien phrases are only a hindrance to science. I will call it simply the ***primary larva***, as I have done in the case of the Meloides, the Leucospis, and the Anthrax.

The form of the primary larva of the Cigale is eminently adapted to its conditions and facilitates its escape. The tunnel in which the egg is hatched is very narrow, leaving only just room for passage. Moreover, the eggs are arranged in a row, not end to end, but partially overlapping. The larva escaping from the hinder ranks has to squeeze past the empty shells, still in position, of the eggs which have already hatched, so that the narrowness of the passage is increased by the empty egg-shells. Under these conditions the larva as it will be presently, when it has torn its temporary wrappings, would be unable to effect the difficult passage. With the encumbrance of antennae, with long limbs spreading far out from the axis of the body, with curved, pointed talons which hook themselves into their medium of support, everything would militate against a prompt liberation. The eggs in one chamber hatch almost simultaneously. It is therefore essential that the first-born larvae should hurry out of their shelter as quickly as possible, leaving the passage free for those behind them. Hence the boat-like shape, the smooth hairless body without projections, which easily squeezes its way past obstructions. The primary larva, with its various appendages closely wrapped against its body by a common sheath, with its fish-like form and its single and only partially movable limb, is perfectly adapted to make the difficult passage to the outer air.

This phase is of short duration. Here, for instance, a migrating larva shows its head, with its big black eyes, and raises the broken fibres of the entrance. It gradually works itself forward, but so slowly that the magnifying-glass scarcely reveals its progress. At the end of half an hour at the shortest we see the entire body of the creature; but the orifice by which it is escaping still holds it by the hinder end of the body.

Then, without further delay, the coat which it wears for this rough piece of work begins to split, and the larva skins itself, coming out of its wrappings head first. It is then the normal larva; the only form known to Reaumur. The rejected coat forms a suspensory thread, expanding at its free end to form a little cup. In this cup is inserted the end of the abdomen of the larva, which, before allowing itself to fall to earth, takes a sun-bath, grows harder, stretches itself, and tries its strength, lightly swinging at the end of its life-line.

This little flea, as Reaumur calls it, first white, then amber-coloured, is precisely the larva which will delve in the earth. The antennae, of fair length, are free and waving to and fro; the limbs are bending at their articulations; the fore-limbs, which are relatively powerful, open and shut their talons. I can scarcely think of any more curious spectacle than that of this tiny gymnast hanging by its tail, swinging to the faintest breath, and preparing in the air for its entry into the world. It hangs there for a variable period; some larvae let themselves fall at the end of half an hour; others spend hours in their long-stemmed cup; some even remain suspended until the following day.

Whether soon or late, the fall of the larva leaves suspended the thread by which it hung, the wrappings of the primary larva. When all the brood have disappeared, the aperture of the nest is thus hung with a branch of fine, short threads, twisted and knotted together, like dried white of egg. Each thread is expanded into a tiny cup at its free end. These are very delicate and ephemeral relics, which perish at a touch. The least wind quickly blows them away.

Let us return to the larva. Sooner or later, as we have seen, it falls to the ground, either by accident or intention. The tiny creature, no bigger than a flea, has preserved its tender newly-hatched flesh from contact with the rough earth by hanging in the air until its tissues have hardened. Now it plunges into the troubles of life.

I foresee a thousand dangers ahead. A mere breath of wind may carry this atom away, and cast it on that inaccessible rock in the midst of a rut in the road which still contains a little water; or on the sand, the region of famine where nothing grows; or upon a soil of clay, too tenacious to be tunnelled. These mortal accidents are frequent, for gusts of wind are frequent in the windy and already severe weather of the end of October.

This delicate organism requires a very soft soil, which can easily be entered, so that it may immediately obtain a suitable shelter. The cold days are coming; soon the frosts will be here. To wander on the surface would expose it to grave perils. It must contrive without delay to descend into the earth, and that to no trivial depth. This is the unique and imperative condition of safety, and in many cases it is impossible of realisation. What use are the claws of this tiny flea against rock, sandstone, or hardened clay? The creature must perish if it cannot find a subterranean refuge in good time.

Everything goes to show that the necessity of this first foothold on the soil, subject as it is to so many accidents, is the cause of the great mortality in the Cigale family. The little black parasite, the destroyer of eggs, in itself evokes the necessity of a large batch of eggs; and the difficulty which the larva experiences in effecting a safe lodgment in the earth is yet another explanation of the fact that the maintenance of the race at its proper strength requires a batch of three or four hundred eggs from each mother. Subject to many accidents, the Cigale is fertile to excess. By the prodigality of her ovaries she conjures the host of perils which threaten her offspring.

During the rest of my experiment I can at least spare the larvae the worst difficulties of their first establishment underground. I take some soil from the heath, which is very soft and almost black, and I pass it through a fine sieve. Its colour will enable me more easily to find the tiny fair-skinned larvae when I wish to inform myself of passing events; its lightness makes it a suitable refuge for such weak and fragile beings. I pack it Pretty firmly in a glass vase; I plant in it a little tuft of thyme; I sow in it a few grains of wheat. There is no hole at the bottom of the vase, although there should be one for the benefit of the thyme and the corn; but the captives would find it and escape by it. The plantation and the crop will suffer from this lack of drainage, but at least I am sure of recovering my larvae with the help of patience and a magnifying-glass. Moreover, I shall go gently in the matter of irrigation, giving only just enough water to save the plants from perishing.

When all is in order, and when the wheat is beginning to shoot, I place six young larvae of the Cigale on the surface of the soil. The tiny creatures begin to pace hither and thither; they soon explore the surface of their world, and some try vainly to climb the sides of the vase. Not one of them seems inclined to bury itself; so that I ask myself anxiously what can be the object of their prolonged and active explorations. Two hours go by, but their wanderings continue.

What do they want? Food? I offer them some tiny bulbs with bundles of sprouting roots, a few fragments of leaves and some fresh blades of grass. Nothing tempts them; nothing brings them to a standstill. Apparently they are seeking for a favourable point before descending into the earth. But there is no need for this hesitating exploration on the soil I have prepared for them; the whole area, or so it seems to me, lends itself excellently to the operations which I am expecting to see them commence. Yet apparently it will not answer the purpose.

Under natural conditions a little wandering might well be indispensable. Spots as soft as my bed of earth from the roots of the briar-heather, purged of all hard bodies and finely sifted, are rare in nature. Coarse soils are more usual, on which the tiny creatures could make no impression. The larva must wander at hazard, must make a pilgrimage of indefinite duration before finding a favourable place. Very many, no doubt, perish, exhausted by their fruitless search. A voyage of exploration in a country a few inches wide evidently forms part of the curriculum of young Cigales. In my glass prison, so luxuriously furnished, this pilgrimage is useless. Never mind: it must be accomplished according to the consecrated rites.

At last my wanderers grow less excited. I see them attack the earth with the curved talons of their fore-limbs, digging their claws into it and making such an excavation as the point of a thick needle would enter. With a magnifying-glass I watch their picks at work. I see their talons raking atom after atom of earth to the surface. In a few minutes there is a little gaping well. The larva climbs downwards and buries itself, henceforth invisible.

On the morrow I turn out the contents of the vase without breaking the mould, which is held together by the roots of the thyme and the wheat. I find all my larvae at the bottom, arrested by the glass. In twenty-four hours they had sunk themselves through the entire thickness of the earth--a matter of some four inches. But for obstacle at the bottom they would have sunk even further.

On the way they have probably encountered the rootlets of my little plantation. Did they halt in order to take a little nourishment by implanting their proboscis? This is hardly probable, for a few rootlets were pressed against the bottom of the glass, but none of my prisoners were feeding. Perhaps the shock of reversing the pot detached them.

It is obvious that underground there is no other nourishment for them than the sap of roots. Adult or larva, the Cigale is a strict vegetarian. As an adult insect it drinks the sap of twigs and branches; as a larva it sucks the sap of roots. But at what stage does it take the first sip? That I do not know as yet, but the foregoing experiment seems to show that the newly hatched larva is in greater haste to burrow deep into the soil, so as to obtain shelter from the coming winter, than to station itself at the roots encountered in its passage downwards.

I replace the mass of soil in the vase, and the six exhumed larvae are once more placed on the surface of the soil. This time they commence to dig at once, and have soon disappeared. Finally the vase is placed in my study window, where it will be subject to the influences, good and ill, of the outer air.

A month later, at the end of November, I pay the young Cigales a second visit. They are crouching, isolated at the bottom of the mould. They do not adhere to the roots; they have not grown; their appearance has not altered. Such as they were at the beginning of the experiment, such they are now, but rather less active. Does not this lack of growth during November, the mildest month of winter, prove that no nourishment is taken until the spring?

The young Sitares, which are also very minute, directly they issue from the egg at the entrance of the tubes of the Anthrophorus, remain motionless, assembled in a heap, and pass the whole of the winter in a state of complete abstinence. The young Cigales apparently behave in a very similar fashion. Once they have burrowed to such depths as will safeguard them from the frosts they sleep in solitude in their winter quarters, and await the return of spring before piercing some neighbouring root and taking their first repast.

I have tried unsuccessfully to confirm these deductions by observation. In April I unpotted my plant of thyme for the third time. I broke up the mould and spread it under the magnifying-glass. It was like looking for needles in a haystack; but at last I recovered my little Cigales. They were dead, perhaps of cold, in spite of the bell-glass with which I had covered the pot, or perhaps of starvation, if the thyme was not a suitable food-plant. I give up the problem as too difficult of solution.

To rear such larvae successfully one would require a deep, extensive bed of earth which would shelter them from the winter cold; and, as I do not know what roots they prefer, a varied vegetation, so that the little creatures could choose according to their taste. These conditions are by no means impracticable, but how, in the large earthy mass, containing at least a cubic yard of soil, should we recover the atoms I had so much trouble to find in a handful of black soil from the heath? Moreover, such a laborious search would certainly detach the larva from its root.

The early subterranean life of the Cigale escapes us. That of the maturer larva is no better known. Nothing is more common, while digging in the fields to any depth, to find these impetuous excavators under the spade; but to surprise them fixed upon the roots which incontestably nourish them is quite another matter. The disturbance of the soil warns the larva of danger. It withdraws its proboscis in order to retreat along its galleries, and when the spade uncovers it has ceased to feed.

If the hazards of field-work, with its inevitable disturbance of the larvae, cannot teach us anything of their subterranean habits, we can at least learn something of the duration of the larval stage. Some obliging farmers, who were making some deep excavations in March, were good enough to collect for me all the larvae, large and small, unearthed in the course of their labour. The total collection amounted to several hundreds. They were divided, by very clearly marked differences of size, into three categories: the large larvae, with rudiments of wings, such as those larvae caught upon leaving the earth possess; the medium-sized, and the small. Each of these stages must correspond to a different age. To these we may add the larvae produced by the last hatching of eggs, creatures too minute to be noticed by my rustic helpers, and we obtain four years as the probable term of the larvae underground.

The length of their aerial existence is more easily computed. I hear the first Cigales about the summer solstice. A month later the orchestra has attained its full power. A very few late singers execute their feeble solos until the middle of September. This is the end of the concert. As all the larvae do not issue from the ground at the same time, it is evident that the singers of September are not contemporary with those that began to sing at the solstice. Taking the average between these two dates, we get five weeks as the probable duration of the Cigales' life on earth.

Four years of hard labour underground, and a month of feasting in the sun; such is the life of the Cigale. Do not let us again reproach the adult insect with his triumphant delirium. For four years, in the darkness he has worn a dirty parchment overall; for four years he has mined the soil with his talons, and now the mud-stained sapper is suddenly clad in the finest raiment, and provided with wings that rival the bird's; moreover, he is drunken with heat and flooded with light, the supreme terrestrial joy. His cymbals will never suffice to celebrate such felicity, so well earned although so ephemeral.

CHAPTER V
THE MANTIS.--THE CHASE

There is another creature of the Midi which is quite as curious and interesting as the Cigale, but much less famous, as it is voiceless. If Providence had provided it with cymbals, which are a prime element of popularity, it would soon have eclipsed the renown of the celebrated singer, so strange is its shape, and so peculiar its manners. It is called by the Provencals *lou Prego-Dieu*, the creature which prays to God. Its official name is the Praying Mantis (Mantis religiosa, Lin.).

For once the language of science and the vocabulary of the peasant agree. Both represent the Mantis as a priestess delivering oracles, or an ascetic in a mystic ecstasy. The comparison is a matter of antiquity. The ancient Greeks called the insect [Greek: Mantis], the divine, the prophet. The worker in the fields is never slow in perceiving analogies; he will always generously supplement the vagueness of the facts. He has seen, on the sun-burned herbage of the meadows, an insect of commanding appearance, drawn up in majestic attitude. He has noticed its wide, delicate wings of green, trailing behind it like long linen veils; he has seen its forelimbs, its arms, so to speak, raised towards to the sky in a gesture of invocation. This was enough: popular imagination has done the rest; so that since the period of classical antiquity the bushes have been peopled with priestesses emitting oracles and nuns in prayer.

Good people, how very far astray your childlike simplicity has led you! These attitudes of prayer conceal the most atrocious habits; these supplicating arms are lethal weapons; these fingers tell no rosaries, but help to exterminate the unfortunate passer-by. It is an exception that we should never look for in the vegetarian family of the Orthoptera, but the Mantis lives exclusively upon living prey. It is the tiger of the peaceful insect peoples; the ogre in ambush which demands a tribute of living flesh. If it only had sufficient strength its blood-thirsty appetites, and its horrible perfection of concealment would make it the terror of the countryside. The **Prego-Dieu** would become a Satanic vampire.

Apart from its lethal weapon the Mantis has nothing about it to inspire apprehension. It does not lack a certain appearance of graciousness, with its slender body, its elegant waist-line, its tender green colouring, and its long gauzy wings. No ferocious jaws, opening like shears; on the contrary, a fine pointed muzzle which seems to be made for billing and cooing. Thanks to a flexible neck, set freely upon the thorax, the head can turn to right or left as on a pivot, bow, or raise itself high in the air. Alone among insects, the Mantis is able to direct its gaze; it inspects and examines; it has almost a physiognomy.

There is a very great contrast between the body as a whole, which has a perfectly peaceable aspect, and the murderous fore-limbs. The haunch of the fore-limb is unusually long and powerful. Its object is to throw forward the living trap which does not wait for the victim, but goes in search of it. The snare is embellished with a certain amount of ornamentation. On the inner face the base of the haunch is decorated with a pretty black spot relieved by smaller spots of white, and a few rows of fine pearly spots complete the ornamentation.

The thigh, still longer, like a flattened spindle, carries on the forward half of the lower face a double row of steely spines. The innermost row contains a dozen, alternately long and black and short and green. This alternation of unequal lengths makes the weapon more effectual for holding. The outer row is simpler, having only four teeth. Finally, three needle-like spikes, the longest of all, rise behind the double series of spikes. In short, the thigh is a saw with two parallel edges, separated by a groove in which the foreleg lies when folded.

The foreleg, which is attached to the thigh by a very flexible articulation, is also a double-edged saw, but the teeth are smaller, more numerous, and closer than those of the thigh. It terminates in a strong hook, the point of which is as sharp as the finest needle: a hook which is fluted underneath and has a double blade like a pruning-knife.

A weapon admirably adapted for piercing and tearing, this hook has sometimes left me with visible remembrances. Caught in turn by the creature which I had just captured, and not having both hands free, I have often been obliged to get a second person to free me from my tenacious captive! To free oneself by violence without disengaging the firmly implanted talons would result in lacerations such as the thorns of a rosebush will produce. None of our insects is so inconvenient to handle. The Mantis digs its knife-blades into your flesh, pierces you with its needles, seizes you as in a vice, and renders self-defence almost impossible if, wishing to take your quarry alive, you refrain from crushing it out of existence.

When the Mantis is in repose its weapons are folded and pressed against the thorax, and are perfectly inoffensive in appearance. The insect is apparently praying. But let a victim come within reach, and the attitude of prayer is promptly abandoned. Suddenly unfolded, the three long joints of the deadly fore-limbs shoot out their terminal talons, which strike the victim and drag it backwards between the two saw-blades of the thighs. The vice closes with a movement like that of the forearm upon the upper arm, and all is over; crickets, grasshoppers, and even more powerful insects, once seized in this trap with its four rows of teeth, are lost irreparably. Their frantic struggles will never release the hold of this terrible engine of destruction.

The habits of the Mantis cannot be continuously studied in the freedom of the fields; the insect must be domesticated. There is no difficulty here; the Mantis is quite indifferent to imprisonment under glass, provided it is well fed. Offer it a tasty diet, feed it daily, and it will feel but little regret for its native thickets.

For cages I use a dozen large covers of wire gauze, such as are used in the larder to protect meat from the flies. Each rests upon a tray full of sand. A dry tuft of thyme and a flat stone on which the eggs may be laid later on complete the furnishing of such a dwelling. These cages are placed in a row on the large table in my entomological laboratory, where the sun shines on them during the greater part of the day. There I install my captives; some singly, some in groups.

It is in the latter half of August that I begin to meet with the adult insect on the faded herbage and the brambles at the roadside. The females, whose bellies are already swollen, are more numerous every day. Their slender companions, on the other hand, are somewhat rare, and I often have some trouble in completing my couples; whose relations will finally be terminated by a tragic consummation. But we will reserve these amenities for a later time, and will consider the females first.

They are tremendous eaters, so that their entertainment, when it lasts for some months is not without difficulties. Their provisions must be renewed every day, for the greater part are disdainfully tasted and thrown aside. On its native bushes I trust the Mantis is more economical. Game is not too abundant, so that she doubtless devours her prey to the last atom; but in my cages it is always at hand. Often, after a few mouthfuls, the insect will drop the juicy morsel without displaying any further interest in it. Such is the ennui of captivity!

To provide them with a luxurious table I have to call in assistants. Two or three of the juvenile unemployed of my neighbourhood, bribed by slices of bread and jam or of melon, search morning and evening on the neighbouring lawns, where they fill their game-bags, little cases made from sections of reeds, with living grasshoppers and crickets. On my own part, I make a daily tour of the paddock, net in hand, with the object of obtaining some choice dish for my guests.

These particular captures are destined to show me just how far the vigour and audacity of the Mantis will lead it. They include the large grey cricket (Pachytylus cinerascens, Fab.), which is larger than the creature which devours it; the white-faced Decticus, armed with powerful mandibles from which it is wise to guard one's fingers; the grotesque Truxalis, wearing a pyramidal mitre on its head; and the Ephippigera of the vineyards, which clashes its cymbals and carries a sabre at the end of its barrel-shaped abdomen. To this assortment of disobliging creatures let us add two horrors: the silky Epeirus, whose disc-shaped scalloped abdomen is as big as a shilling, and the crowned Epeirus, which is horribly hairy and corpulent.

I cannot doubt that the Mantis attacks such adversaries in a state of nature when I see it, under my wire-gauze covers, boldly give battle to whatever is placed before it. Lying in wait among the bushes it must profit by the prizes bestowed upon it by hazard, as in its cage it profits by the wealth of diet due to my generosity. The hunting of such big game as I offer, which is full of danger, must form part of the creature's usual life, though it may be only an occasional pastime, perhaps to the great regret of the Mantis.

Crickets of all kinds, butterflies, bees, large flies of many species, and other insects of moderate size: such is the prey that we habitually find in the embrace of the murderous arms of the Mantis. But in my cages I have never known the audacious huntress to recoil before any other insect. Grey cricket, Decticus, Epeirus or

Truxalis, sooner or later all are harpooned, held motionless between the saw-edges of the arms, and deliciously crunched at leisure. The process deserves a detailed description.

At the sight of a great cricket, which thoughtlessly approaches along the wire-work of the cover, the Mantis, shaken by a convulsive start, suddenly assumes a most terrifying posture. An electric shock would not produce a more immediate result. The transition is so sudden, the mimicry so threatening, that the unaccustomed observer will draw back his hand, as though at some unknown danger. Seasoned as I am, I myself must confess to being startled on occasions when my thoughts have been elsewhere. The creature spreads out like a fan actuated by a spring, or a fantastic Jack-in-the-box.

The wing-covers open, and are thrust obliquely aside; the wings spring to their full width, standing up like parallel screens of transparent gauze, forming a pyramidal prominence which dominates the back; the end of the abdomen curls upwards crosier-wise, then falls and unbends itself with a sort of swishing noise, a ***pouf! pouf!*** like the sound emitted by the feathers of a strutting turkey-cock. One is reminded of the puffing of a startled adder.

Proudly straddling on its four hind-claws, the insect holds its long body almost vertical. The murderous fore-limbs, at first folded and pressed against one another on the thorax, open to their full extent, forming a cross with the body, and exhibiting the axillae ornamented with rows of pearls, and a black spot with a central point of white. These two eyes, faintly recalling those of the peacock's tail, and the fine ebony embossments, are part of the blazonry of conflict, concealed upon ordinary occasions. Their jewels are only assumed when they make themselves terrible and superb for battle.

Motionless in its weird position, the Mantis surveys the acridian, its gaze fixed upon it, its head turning gently as on a pivot as the other changes place. The object of this mimicry seems evident; the Mantis wishes to terrorise its powerful prey, to paralyse it with fright; for if not demoralised by fear the quarry might prove too dangerous.

Does it really terrify its prey? Under the shining head of the Decticus, behind the long face of the cricket, who is to say what is passing? No sign of emotion can reveal itself upon these immovable masks. Yet it seems certain that the threatened creature is aware of its danger. It sees, springing up before it, a terrible spectral form with talons outstretched, ready to fall upon it; it feels itself face to face with death, and fails to flee while yet there is time. The creature that excels in leaping, and might so easily escape from the threatening claws, the wonderful jumper with the prodigious thighs, remains crouching stupidly in its place, or even approaches the enemy with deliberate steps.[2]

It is said that young birds, paralysed with terror by the gaping mouth of a serpent, or fascinated by its gaze, will allow themselves to be snatched from the nest, incapable of movement. The cricket will often behave in almost the same way. Once within reach of the enchantress, the grappling-hooks are thrown, the fangs strike, the double saws close together and hold the victim in a vice. Vainly the captive struggles; his mandibles chew the air, his desperate kicks meet with no resistance. He has met with his fate. The Mantis refolds her wings, the standard of battle; she resumes her normal pose, and the meal commences.

In attacking the Truxalis and the Ephippigera, less dangerous game than the grey cricket and the Decticus, the spectral pose is less imposing and of shorter duration. It is often enough to throw forward the talons; this is so in the case of the Epeirus, which is seized by the middle of the body, without a thought of its venomous claws. With the smaller crickets, which are the customary diet in my cages as at liberty, the Mantis rarely employs her means of intimidation; she merely seizes the heedless passer-by as she lies in wait.

When the insect to be captured may present some serious resistance, the Mantis is thus equipped with a pose which terrifies or perplexes, fascinates or absorbs the prey, while it enables her talons to strike with greater certainty. Her gins close on a demoralised victim, incapable of or unready for defence. She freezes the quarry with fear or amazement by suddenly assuming the attitude of a spectre.

The wings play an important part in this fantastic pose. They are very wide, green on the outer edge, but colourless and transparent elsewhere. Numerous nervures, spreading out fan-wise, cross them in the direction of their length. Others, transversal but finer, cut the first at right angles, forming with them a multitude of meshes. In the spectral attitude the wings are outspread and erected in two parallel planes which are almost in contact, like the wings of butterflies in repose. Between the two the end of the abdomen rapidly curls and uncurls. From the rubbing of the belly against the network of nervures proceeds the species of puffing sound which I have compared to the hissing of an adder in a posture of defence. To imitate this curious sound it is enough rapidly to stroke the upper face of an outstretched wing with the tip of the finger-nail.

In a moment of hunger, after a fast of some days, the large grey cricket, which is as large as the Mantis or larger, will be entirely consumed with the exception of the wings, which are too dry. Two hours are sufficient for the completion of this enormous meal. Such an orgy is rare. I have witnessed it two or three times, always asking myself where the gluttonous creature found room for so much food, and how it contrived to reverse in its own favour the axiom that the content is less than that which contains it. I can only admire the privileges of a stomach in which matter is digested immediately upon entrance, dissolved and made away with.

The usual diet of the Mantis under my wire cages consists of crickets of different species and varying greatly in size. It is interesting to watch the Mantis nibbling at its cricket, which it holds in the vice formed by its murderous fore-limbs. In spite of the fine-pointed muzzle, which hardly seems made for such ferocity, the entire insect disappears excepting the wings, of which only the base, which is slightly fleshy, is consumed. Legs, claws, horny integuments, all else is eaten. Sometimes the great hinder thigh is seized by the knuckle, carried to the mouth, tasted, and crunched with a little air of satisfaction. The swollen thigh of the cricket might well be a choice "cut" for the Mantis, as a leg of lamb is for us!

The attack on the victim begins at the back of the neck or base of the head. While one of the murderous talons holds the quarry gripped by the middle of the body, the other presses the head downwards, so that the articulation between the back and the neck is stretched and opens slightly. The snout of the Mantis gnaws and burrows into this undefended spot with a certain persistence, and a large wound is opened in the neck. At the lesion of the cephalic ganglions the struggles of the cricket grow less, and the victim becomes a motionless corpse. Thence, unrestricted in its movements, this beast of prey chooses its mouthfuls at leisure.

CHAPTER VI
THE MANTIS.--COURTSHIP

The little we have seen of the customs of the Mantis does not square very well with the popular name for the insect. From the term *Prego-Dieu* we should expect a peaceful placid creature, devoutly self-absorbed; and we find a cannibal, a ferocious spectre, biting open the heads of its captives after demoralising them with terror. But we have yet to learn the worst. The customs of the Mantis in connection with its own kin are more atrocious even than those of the spiders, who bear an ill repute in this respect.

To reduce the number of cages on my big laboratory table, to give myself a little more room, while still maintaining a respectable menagerie, I installed several females under one cover. There was sufficient space in the common lodging and room for the captives to move about, though for that matter they are not fond of movement, being heavy in the abdomen. Crouching motionless against the wire work of the cover, they will digest their food or await a passing victim. They lived, in short, just as they lived on their native bushes.

Communal life has its dangers. When the hay is low in the manger donkeys grow quarrelsome, although usually so pacific. My guests might well, in a season of dearth, have lost their tempers and begun to fight one another; but I was careful to keep the cages well provided with crickets, which were renewed twice a day. If civil war broke out famine could not be urged in excuse.

At the outset matters did not go badly. The company lived in peace, each Mantis pouncing upon and eating whatever came her way, without interfering with her neighbours. But this period of concord was of brief duration. The bellies of the insects grew fuller: the eggs ripened in their ovaries: the time of courtship and the laying season was approaching. Then a kind of jealous rage seized the females, although no male was present to arouse such feminine rivalry. The swelling of the ovaries perverted my flock, and infected them with an insane desire to devour one another. There were threats, horrid encounters, and cannibal feasts. Once more the spectral pose was seen, the hissing of the wings, and the terrible gesture of the talons outstretched and raised above the head. The females could not have looked more terrible before a grey cricket or a Decticus. Without any motives that I could see, two neighbours suddenly arose in the attitude of conflict. They turned their heads to the right and the left, provoking one another, insulting one another. The *pouf! pouf!* of the wings rubbed by the abdomen sounded the charge. Although the duel was to terminate at the first scratch, without any more serious consequence, the murderous talons, at first folded, open like the leaves of a book, and are extended laterally to protect the long waist and abdomen. The pose is superb, but less terrific than that assumed when the fight is to be to the death.

Then one of the grappling-hooks with a sudden spring flies out and strikes the rival; with the same suddenness it flies back and assumes a position of guard. The adversary replies with a riposte. The fencing reminds one not a little of two cats boxing one another's ears. At the first sign of blood on the soft abdomen, or even at the slightest wound, one admits herself to be conquered and retires. The other refurls her battle standard and goes elsewhere to meditate the capture of a cricket, apparently calm, but in reality ready to recommence the quarrel.

Very often the matter turns out more tragically. In duels to the death the pose of attack is assumed in all its beauty. The murderous talons unfold and rise in the air. Woe to the vanquished! for the victor seizes her in her vice-like grip and at once commences to eat her; beginning, needless to say, at the back of the neck. The odious meal proceeds as calmly as if it were merely a matter of munching a grasshopper; and the survivor enjoys her sister quite as much as lawful game. The spectators do not protest, being only too willing to do the like on the first occasion.

Ferocious creatures! It is said that even wolves do not eat one another. The Mantis is not so scrupulous; she will eat her fellows when her favourite quarry, the cricket, is attainable and abundant.

These observations reach a yet more revolting extreme. Let us inquire into the habits of the insect at breeding time, and to avoid the confusion of a crowd let us isolate the couples under different covers. Thus each pair will have their own dwelling, where nothing can trouble their honeymoon. We will not forget to provide them with abundant food; there shall not be the excuse of hunger for what is to follow.

We are near the end of August. The male Mantis, a slender and elegant lover, judges the time to be propitious. He makes eyes at his powerful companion; he turns his head towards her; he bows his neck and raises his thorax. His little pointed face almost seems to wear an expression. For a long time he stands thus motionless, in contemplation of the desired one. The latter, as though indifferent, does not stir. Yet the lover has seized upon a sign of consent: a sign of which I do not know the secret. He approaches: suddenly he erects his wings, which are shaken with a convulsive tremor.

This is his declaration. He throws himself timidly on the back of his corpulent companion; he clings to her desperately, and steadies himself. The prelude to the embrace is generally lengthy, and the embrace will sometimes last for five or six hours.

Nothing worthy of notice occurs during this time. Finally the two separate, but they are soon to be made one flesh in a much more intimate fashion. If the poor lover is loved by his mistress as the giver of fertility, she also loves him as the choicest of game. During the day, or at latest on the morrow, he is seized by his companion, who first gnaws through the back of his neck, according to use and wont, and then methodically devours him, mouthful by mouthful, leaving only the wings. Here we have no case of jealousy, but simply a depraved taste.

I had the curiosity to wonder how a second male would be received by a newly fecundated female. The result of my inquiry was scandalous. The Mantis in only too many cases is never sated with embraces and conjugal feasts. After a rest, of variable duration, whether the eggs have been laid or not, a second male is welcomed and devoured like the first. A third succeeds him, does his duty, and affords yet another meal. A fourth suffers a like fate. In the course of two weeks I have seen the same Mantis treat seven husbands in this fashion. She admitted all to her embraces, and all paid for the nuptial ecstasy with their lives.

There are exceptions, but such orgies are frequent. On very hot days, when the atmospheric tension is high, they are almost the general rule. At such times the Mantis is all nerves. Under covers which contain large households the females devour one another more frequently than ever; under the covers which contain isolated couples the males are devoured more eagerly than usual when their office has been fulfilled.

I might urge, in mitigation of these conjugal atrocities, that the Mantis does not commit them when at liberty. The male, his function once fulfilled, surely has time to wander off, to escape far away, to flee the terrible spouse, for in my cages he is given a respite, often of a whole day. What really happens by the roadside and in the thickets I do not know; chance, a poor schoolmistress, has never instructed me concerning the love-affairs of the Mantis when at liberty. I am obliged to watch events in my laboratory, where the captives, enjoying plenty of sunshine, well nourished, and comfortably lodged, do not seem in any way to suffer from nostalgia. They should behave there as they behave under normal conditions.

Alas! the facts force me to reject the statement that the males have time to escape; for I once surprised a male, apparently in the performance of his vital functions, holding the female tightly embraced--but he had no head, no neck, scarcely any thorax! The female, her head turned over her shoulder, was peacefully browsing on the remains of her lover! And the masculine remnant, firmly anchored, continued its duty!

Love, it is said, is stronger than death! Taken literally, never has an aphorism received a more striking confirmation. Here was a creature decapitated, amputated as far as the middle of the thorax; a corpse which still struggled to give life. It would not relax its hold until the abdomen itself, the seat of the organs of procreation, was attacked.

The custom of eating the lover after the consummation of the nuptials, of making a meal of the exhausted pigmy, who is henceforth good for nothing, is not so difficult to understand, since insects can hardly be accused of sentimentality; but to devour him during the act surpasses anything that the most morbid mind could imagine. I have seen the thing with my own eyes, and I have not yet recovered from my surprise.

Could this unfortunate creature have fled and saved himself, being thus attacked in the performance of his functions? No. We must conclude that the loves of the Mantis are fully as tragic, perhaps even more so, than those of the spider. I do not deny that the limited area of the cage may favour the massacre of the males; but the cause of such butchering must be sought elsewhere. It is perhaps a reminiscence of the carboniferous period when the insect world gradually took shape through prodigious procreation. The Orthoptera, of which the Mantes form a branch, are the first-born of the insect world.

Uncouth, incomplete in their transformation, they wandered amidst the arborescent foliage, already flourishing when none of the insects sprung of more complex forms of metamorphosis were as yet in existence: neither butterflies, beetles, flies, nor bees. Manners were not gentle in those epochs, which were full of the lust to destroy in order to produce; and the Mantis, a feeble memory of those ancient ghosts, might well preserve the customs of an earlier age. The utilisation of the males as food is a custom in the case of other members of the Mantis family. It is, I must admit, a general habit. The little grey Mantis, so small and looking so harmless in her cage, which never seeks to harm her neighbours in spite of her crowded quarters, falls upon her male and devours him as ferociously as the Praying Mantis. I have worn myself out in trying to procure the indispensable complements to my female specimens. No sooner is my capture, strongly winged, vigorous and alert, introduced into the cage than he is seized, more often than not, by one of the females who no longer have need of his assistance and devoured. Once the ovaries are satisfied the two species of Mantis conceive an antipathy for the male; or rather they regard him merely as a particularly tasty species of game.

CHAPTER VII
THE MANTIS.--THE NEST

L et us take a more pleasant aspect of the insect whose loves are so tragic. Its nest is a marvel. In scientific language it is known as the *ootek*, or the "egg-box." I shall not make use of this barbarous expression. As one does not speak of the "egg-box" of the titmouse, meaning "the nest of the titmouse," why should I invoke the box in speaking of the Mantis? It may look more scientific; but that does not interest me.

The nest of the Praying Mantis may be found almost everywhere in places exposed to the sun: on stones, wood, vine stocks, the twigs of bushes, stems of dried grass, and even on products of human industry, such as fragments of brick, rags of heavy cloth, and pieces of old boots. Any support will suffice, so long as it offers inequalities to which the base of the nest may adhere, and so provide a solid foundation. The usual dimensions of the nest are one and a half inches long by three-quarters of an inch wide, or a trifle larger. The colour is a pale tan, like that of a grain of wheat. Brought in contact with a flame the nest burns readily, and emits an odour like that of burning silk. The material of the nest is in fact a substance similar to silk, but instead of being drawn into a thread it is allowed to harden while a mass of spongy foam. If the nest is fixed on a branch the base creeps round it, envelops the neighbouring twigs, and assumes a variable shape according to the accidents of support; if it is fixed on a flat surface the under side, which is always moulded by the support, is itself flat. The nest then takes the form of a demi-ellipsoid, or, in other words, half an egg cut longitudinally; more or less obtuse at one end, but pointed at the other, and sometimes ending in a short curved tail.

In all cases the upper face is convex and regular. In it we can distinguish three well-marked and longitudinal zones. The middle zone, which is narrower than the others, is composed of thin plates arranged in couples, and overlapping like the tiles of a roof. The edges of these plates are free, leaving two parallel series of fissures by which the young can issue when the eggs are hatched. In a nest recently abandoned this zone is covered with fine cast-off skins which shiver at the least breath, and soon disappear when exposed to the open air. I will call this zone the zone of issue, as it is only along this bell that the young can escape, being set free by those that have preceded them.

In all other directions the cradle of this numerous family presents an unbroken wall. The two lateral zones, which occupy the greater part of the demi-ellipsoid, have a perfect continuity of surface. The little Mantes, which are very feeble when first hatched, could not possibly make their way through the tenacious substance of the walls. On the interior of these walls are a number of fine transverse furrows, signs of the various layers in which the mass of eggs is disposed.

Let us cut the nest in half transversely. We shall then see that the mass of eggs constitutes an elongated core, of very firm consistency, surrounded as to the bottom and sides by a thick porous rind, like solidified foam. Above the eggs are the curved plates, which are set very closely and have little freedom; their edges constituting the zone of issue, where they form a double series of small overlapping scales.

The eggs are set in a yellowish medium of horny appearance. They are arranged in layers, in lines forming arcs of a circle, with the cephalic extremities converging towards the zone of issue. This orientation tells us of the method of delivery. The newly-born larvae will slip into the interval between two adjacent flaps or leaves, which form a prolongation of the core; they will then find a narrow passage, none too easy to effect, but sufficient, having regard to the curious provision which we shall deal with directly; they will then reach the zone of issue. There, under the overlapping scales, two passages of exit open for each layer of eggs. Half the larvae will issue by the right-hand passage, half by that on the left hand. This process is repeated for each layer, from end to end of the nest.

Let us sum up those structural details, which are not easily grasped unless one has the nest before one. Lying along the axis of the nest, and in shape like a date-stone, is the mass of eggs, grouped in layers. A protective rind, a kind of solidified foam, envelops this core, except at the top, along the central line, where the porous rind is replaced by thin overlapping leaves. The free edges of these leaves form the exterior of the zone of issue; they overlap one another, forming two series of scales, leaving two exits, in the shape of narrow crevices, for each layer of eggs.

To be present at the construction of the nest--to learn how the Mantis contrives to build so complex a structure--such was the main point of my researches. I succeeded, not without difficulty, as the eggs are laid without warning and nearly always at night. After a great deal of futile endeavour, chance at last favoured me. On the 5th of September one of my guests, fecundated on the 29th of August, began to make her preparations under my eyes, at four o'clock in the afternoon.

One remark before proceeding: all the nests I have obtained in the laboratory--and I have obtained a good number--have without exception been built upon the wire gauze of the covers. I have been careful to provide the insects with roughened stones and tufts of thyme, both being very commonly used as foundations in the open fields. The captives have always preferred the network of wire gauze, which affords a perfectly firm foundation, as the soft material of the nest becomes incrusted upon the meshes as it hardens.

In natural conditions the nests are never in any way sheltered; they support the inclemencies of winter, resist rain, wind, frost, and snow, without becoming detached. It is true that the female always selects an uneven support on which the foundations of the nest can be shaped, thus obtaining a firm hold. The site chosen is always the best obtainable within reach, and the wire gauze is constantly adopted as the best foundation obtainable in the cages.

The only Mantis that I was able to observe at the moment of laying her eggs worked upside-down, clinging to the wire near the top of the cover. My presence, my magnifying-glass, my investigations did not disturb her in the least, so absorbed was she in her labours. I was able to lift up the dome of wire gauze, tilt it, reverse it, turn it over and reverse it again, without causing the insect to delay her task for a moment. I was able, with my tweezers, to raise the long wings in order to observe rather more closely what was taking place beneath them; the Mantis took absolutely no notice of me. So far all was well; the female did not move, and lent herself impassively to all the indiscretions of the observer. Nevertheless, matters did not proceed as I had wished, so rapid was the operation and so difficult observation.

The end of the abdomen is constantly immersed in a blob of foam, which does not allow one to grasp the details of the process very clearly. This foam is of a greyish white, slightly viscous, and almost like soapsuds. At the moment of its appearance it adheres slightly to the end of a straw plunged into it. Two minutes later it is solidified and no longer adheres to the straw. In a short time its consistency is that of the substance of an old nest.

The foamy mass consists chiefly of air imprisoned in minute bubbles. This air, which gives the nest a volume very much greater than that of the abdomen of the Mantis, evidently does not issue from the insect although the foam appears at the orifice of the genital organs; it is borrowed from the atmosphere. The Mantis builds more especially with air, which is eminently adapted to protect the nest against changes of temperature. She emits a glutinous substance like the liquid secretion of silk-worms, and with this composition, mixed instantaneously with the outer air, she produces the foam of which the nest is constructed.

She whips the secretion as we whip white of egg, in order to make it rise and stiffen. The extremity of the abdomen opens in a long cleft, forming two lateral ladles which open and shut with a rapid, incessant movement, beating the viscous liquid and converting it into foam as it is secreted. Beside the two oscillating ladles we see the internal organs rising and falling, protruding and retreating like a piston-rod, but it is impossible to observe the precise nature of their action, bathed as they are in the opaque blob of foam.

The end of the abdomen, continually palpitating, rapidly closing and opening its valves, oscillates right and left like a pendulum. From each of these oscillations results a layer of eggs in the interior, and a transversal crevice on the exterior. As it advances in the arc described, suddenly, and at frequent intervals, it plunges deeper into the foam, as though burying something at the bottom of the frothy mass. Each time it does so an egg is doubtless deposited; but the operation is so rapid, and takes place under conditions so unfavourable for observation, that I have never once been enabled to see the oviduct at work. I can only judge of the advent of the eggs by the movements of the end of the abdomen, which is immersed more deeply with a sudden plunging movement.

At the same time the viscous composition is emitted in intermittent waves, and is beaten into a foam by the terminal valves. The foam thus obtained spreads itself over the sides and at the base of the layer of eggs, and projects through the meshes of the wire gauze as a result of the pressure of the abdomen. Thus the spongy envelope is progressively created as the ovaries are gradually emptied.

I imagine, although I cannot speak as the result of direct observation, that for the central core, where the eggs are surrounded by a material more homogeneous than that of the outer shell, the Mantis must employ her secretion as it emerges, without beating it into a foam. The layer of eggs once deposited, the two valves would produce the foam required to envelop the eggs. It is extremely difficult, however, to guess what occurs beneath the veil of foam-like secretion.

In a recent nest the zone of issue is surrounded by a layer of finely porous matter, of a pure matt, almost chalky white, which contrasts distinctly with the remainder of the nest, which is of a dirty white. It resembles the icing composition made by confectioners with whipped white of egg, sugar, and starch, for the ornamentation of cakes.

This snowy border is easily crumbled and easily detached. When it disappears the zone of issue is clearly defined, with its double series of leaves with free edges. Exposure to the weather, wind, and rain result in its disappearance, fragment by fragment, so that old nests preserve no trace of it.

At first sight one is tempted to regard this snowy substance as of a different material to the rest of the nest. But does the Mantis really employ two secretions? No. Anatomy, in the first place, assures us of the unity of the materials of the nest. The organ which secretes the substance of the nest consists of cylindrical tubes, having a curious tangled appearance, which are arranged in two groups of twenty each. They are all filled with a colourless, viscous fluid, which is precisely similar in appearance in all parts of the organ. There is no indication of any organ or secretion which could produce a chalky coloration.

Moreover, the method by which the snowy band is formed rejects the idea of a different material. We see the two caudal appendices of the Mantis sweeping the surface of the foamy mass, and skimming, so to speak, the cream of the cream, gathering it together, and retaining it along the hump of the nest in such a way as to form a band like a ribbon of icing. What remains after this scouring process, or what oozes from the band before it has set, spreads over the sides of the nest in a thin layer of bubbles so fine that they cannot be distinguished without the aid of a lens.

We often see a torrent of muddy water, full of clay in suspension, covered with great streaks and masses of foam. On this fundamental foam, so to call it, which is soiled with earthy matters, we see here and there masses of a beautiful white foam, in which the bubbles are much smaller. A process of selection results from variations in density, and here and there we see foam white as snow resting on the dirty foam from which it is produced. Something of the kind occurs when the Mantis builds her nest. The two appendices whip the viscous secretion of the glands into foam. The lightest portion, whose bubbles are of the greatest tenuity, which is white on account of its finer porosity, rises to the surface, where the caudal filaments sweep it up and gather it into the snowy ribbon which runs along the summit of the nest.

So far, with a little patience, observation is possible and yields a satisfactory result. It becomes impossible in the matter of the complex central zone, where the exits for the larvae are contrived through the double series of overlapping leaves. The little I have been able to learn amounts to this: The end of the abdomen, deeply cleft in a horizontal direction, forms a kind of fork, of which the upper extremity remains almost motionless, while the lower continuously oscillates, producing the foam and depositing the eggs. The creation of the central zone is certainly the work of the upper extremity.

It is always to be seen in the continuation of this central zone, in the midst of the fine white foam gathered up by the caudal filaments. The latter delimit the zone, one working on either side, feeling the edges of the belt, and apparently testing it and judging its progress. These two filaments are like two long fingers of exquisite sensitiveness, which direct the difficult operation.

But how are the two series of scales obtained, and the fissures, the gates of exit which they shelter? I do not know; I cannot even imagine. I leave the end of the problem to others.

What a wonderful mechanism is this, that has the power to emit and to form, so quickly and methodically, the horny medium of the central kernel, the foam which forms the protective walls, the white creamy foam of the ribbon which runs along the central zone, the eggs, and the fecundating liquid, while at the same time it constructs the overlapping leaves, the imbricated scales, and the alternating series of open fissures! We are lost in the face of such a wonder. Yet how easily the work is performed! Clinging to the wire gauze, forming, so to speak, the axis of her nest, the Mantis barely moves. She bestows not a glance on the marvel which is growing behind her; her limbs are used only for support; they take no part in the building of the nest. The nest is built, if we may say so, automatically. It is not the result of industry and the cunning of instinct; it is a purely mechanical task, which is conditioned by the implements, by the organisation of the insect. The nest, complex though it is in structure, results solely from the functioning of the organs, as in our human industries a host of objects are mechanically fashioned whose perfection puts the dexterity of the fingers to shame.

From another point of view the nest of the Mantis is even more remarkable. It forms an excellent application of one of the most valuable lessons of physical science in the matter of the conservation of heat. The Mantis has outstripped humanity in her knowledge of thermic nonconductors or insulators.

The famous physicist Rumford was responsible for a very pretty experiment designed to demonstrate the low conductivity of air where heat other than radiant heat is concerned. The famous scientist surrounded a frozen cheese by a mass of foam consisting of well-beaten eggs. The whole was exposed to the heat of an oven. In a few minutes a light omelette was obtained, piping hot, but the cheese in the centre was as cold as at the outset. The air imprisoned in the bubbles of the surrounding froth accounts for the phenomenon. Extremely refractory to heat, it had absorbed the heat of the oven and had prevented it from reaching the frozen substance in the centre of the omelette.

Now, what does the Mantis do? Precisely what Rumford did; she whips her albumen to obtain a soufflee, a froth composed of myriads of tiny air-bubbles, which will protect the germs of life contained in the central core. It is true that her aim is reversed; the coagulated foam of the nest is a safeguard against cold, not against heat, but what will afford protection from the one will afford protection from the other; so that Rumford, had he wished, might equally well have maintained a hot body at a high temperature in a refrigerator.

Rumford understood the athermic properties of a blanket of air-cells, thanks to the accumulated knowledge of his predecessors and his own studies and experiments. How is it that the Mantis, for who knows how many ages, has been able to outstrip our physicists in this problem in calorics? How did she learn to surround her eggs with this mass of solidifying froth, so that it was able, although fixed to a bough or a stone without other shelter, to brave with impunity the rigours of winter?

The other Mantes found in my neighbourhood, which are the only species of which I can speak with full knowledge, employ or omit the envelope of solidifying froth accordingly as the eggs are or are not intended to survive the winter. The little Grey Mantis (Ameles decolor), which differs so widely from the Praying Mantis in that the wings of the female are almost completely absent, builds a nest hardly as large as a cherry-stone, and covers it skilfully with a porous rind. Why this cellular envelope? Because the nest of the *Ameles*, like that of the Praying Mantis, has to endure through the winter, fixed to a stone or a twig, and is thus exposed to the full severity of the dangerous season.

The *Empusa pauperata*, on the other hand (one of the strangest of European insects), builds a nest as small as that of the *Ameles*, although the insect itself is as large as the Praying Mantis. This nest is quite a small structure, composed of a small number of cells, arranged side by side in three or four series, sloping together at the neck. Here there is a complete absence of the porous envelope, although the nest is exposed to the weather, like the previous examples, affixed to some twig or fragment of rock. The lack of the insulating rind is a sign of different climatic conditions. The eggs of the *Empusa* hatch shortly after they are laid, in warm and sunny weather. Not being exposed to the asperities of the winter, they need no protection other than the thin egg-cases themselves.

Are these nice and reasonable precautions, which rival the experiment of Rumford, a fortuitous result?--one of the innumerable combinations which fall from the urn of chance? If so, let us not recoil before the absurd: let us allow that the blindness of chance is gifted with marvellous foresight.

The Praying Mantis commences her nest at the blunter extremity, and completes it at the pointed tail. The latter is often prolonged in a sort of promontory, in which the insect expends the last drop of glutinous liquid as she stretches herself after her task. A sitting of two hours, more or less, without interruption, is required for the total accomplishment of the work. Directly the period of labour is over, the mother withdraws, indifferent henceforth to her completed task. I have watched her, half expecting to see her return, to discover some tenderness for the cradle of her family. But no: not a trace of maternal pleasure. The work is done, and concerns her no longer. Crickets approach; one of them even squats upon the nest. The Mantis takes no notice of them. They are peaceful intruders, to be sure; but even were they dangerous, did they threaten to rifle the nest, would she attack them and drive them away? Her impassive demeanour convinces me that she would not. What is the nest to her? She is no longer conscious of it.

I have spoken of the many embraces to which the Praying Mantis submits, and of the tragic end of the male, who is almost invariably devoured as though a lawful prey. In the space of a fortnight I have known the same female to adventure upon matrimony no less than seven times. Each time the readily consoled widow devoured her mate. Such habits point to frequent laying; and we find the appearance confirmed, though not as a general rule. Some of my females gave me one nest only; others two, the second as capacious as the first. The most fruitful of all produced three; of these the two first were of normal dimensions, while the third was about half the usual size.

From this we can reckon the productivity of the insect's ovaries. From the transverse fissures of the median zone of the nest it is easy to estimate the layers of eggs; but these layers contain more or fewer eggs according to their position in the middle of the nest or near the ends. The numbers contained by the widest and narrowest layers will give us an approximate average. I find that a nest of fair size contains about four hundred eggs. Thus the maker of the three nests, of which the last was half as large as the others, produced no less than a thousand eggs; eight hundred were deposited in the larger nests and two or three hundred in the smaller. Truly a fine family, but a thought ungainly, were it not that only a few of its members can survive.

Of a fair size, of curious structure, and well in evidence on its twig or stone, the nest of the Praying Mantis could hardly escape the attention of the Provencal peasant. It is well known in the country districts, where it goes by the name of *tigno*; it even enjoys a certain celebrity. But no one seems to be aware of its origin. It is always a surprise to my rustic neighbours when they learn that the well-known *tigno* is the nest of the common Mantis, the *Prego-Dieu*. This ignorance may well proceed from the nocturnal habits of the Mantis. No one has caught the insect at work upon her nest in the silence of the night. The link between the artificer and the work is missing, although both are well known to the villager.

No matter: the singular object exists; it catches the eye, it attracts attention. It must therefore be good for something; it must possess virtue of some kind. So in all ages have the simple reasoned, in the childlike hope of finding in the unfamiliar an alleviation of their sorrows.

By general agreement the rural pharmacopoeia of Provence pronounces the *tigno* to be the best of remedies against chilblains. The method of employment is of the simplest. The nest is cut in two, squeezed and the affected part is rubbed with the cut surface as the juices flow from it. This specific, I am told, is sovereign. All sufferers from blue and swollen fingers should without fail, according to traditional usage, have recourse to the *tigno*.

Is it really efficacious? Despite the general belief, I venture to doubt it, after fruitless experiments on my own fingers and those of other members of my household during the winter of 1895, when the severe and persistent cold produced an abundant crop of chilblains. None of us, treated with the celebrated unguent, observed the swelling to diminish; none of us found that the pain and discomfort was in the least assuaged by the sticky varnish formed by the juices of the crushed *tigno*. It is not easy to believe that others are more successful, but the popular renown of the specific survives in spite of all, probably thanks to a simple accident of identity between the name of the remedy and that of the infirmity: the Provencal for "chilblain" is *tigno*. From the moment when the chilblain and the nest of the Mantis were known by the same name were not the virtues of the latter obvious? So are reputations created.

In my own village, and doubtless to some extent throughout the Midi, the *tigno*--the nest of the Mantis, not the chilblain--is also reputed as a marvellous cure for toothache. It is enough to carry it upon the person to be free of that lamentable affection. Women wise in such matters gather them beneath a propitious moon, and preserve them piously in some corner of the clothes-press or wardrobe. They sew them in the lining of the pocket, lest they should be pulled out with the handkerchief and lost; they will grant the loan of them to a neighbour tormented by some refractory molar. "Lend me thy *tigno*: I am suffering martyrdom!" begs the owner of a swollen face.--"Don't on any account lose it!" says the lender: "I haven't another, and we aren't at the right time of moon!"

We will not laugh at the credulous victim; many a remedy triumphantly puffed on the latter pages of the newspapers and magazines is no more effectual. Moreover, this rural simplicity is surpassed by certain old books which form the tomb of the science of a past age. An English naturalist of the sixteenth century, the well-known physician, Thomas Moffat, informs us that children lost in the country would inquire their way of the Mantis. The insect consulted would extend a limb, indicating the direction to be taken, and, says the author, scarcely ever was the insect mistaken. This pretty story is told in Latin, with an adorable simplicity.

CHAPTER VIII
THE GOLDEN GARDENER.--ITS NUTRIMENT

In writing the first lines of this chapter I am reminded of the slaughter-pens of Chicago; of those horrible meat factories which in the course of the year cut up one million and eighty thousand bullocks and seventeen hundred thousand swine, which enter a train of machinery alive and issue transformed into cans of preserved meat, sausages, lard, and rolled hams. I am reminded of these establishments because the beetle I am about to speak of will show us a compatible celerity of butchery.

In a spacious, glazed insectorium I have twenty-five Carabi aurati. At present they are motionless, lying beneath a piece of board which I gave them for shelter. Their bellies cooled by the sand, their backs warmed by the board, which is visited by the sun, they slumber and digest their food. By good luck I chance upon a procession of pine-caterpillars, in process of descending from their tree in search of a spot suitable for burial, the prelude to the phase of the subterranean chrysalis. Here is an excellent flock for the slaughter-house of the Carabi.

I capture them and place them in the insectorium. The procession is quickly re-formed; the caterpillars, to the number of perhaps a hundred and fifty, move forward in an undulating line. They pass near the piece of board, one following the other like the pigs at Chicago. The moment is propitious. I cry Havoc! and let loose the dogs of war: that is to say, I remove the plank.

The sleepers immediately awake, scenting the abundant prey. One of them runs forward; three, four, follow; the whole assembly is aroused; those who are buried emerge; the whole band of cut-throats falls upon the passing flock. It is a sight never to be forgotten. The mandibles of the beetles are at work in all directions; the procession is attacked in the van, in the rear, in the centre; the victims are wounded on the back or the belly at random. The furry skins are gaping with wounds; their contents escape in knots of entrails, bright green with their aliment, the needles of the pine-tree; the caterpillars writhe, struggling with loop-like movements, gripping the sand with their feet, dribbling and gnashing their mandibles. Those as yet unwounded are digging desperately in the attempt to take refuge underground. Not one succeeds. They are scarcely half buried before some beetle runs to them and destroys them by an eviscerating wound.

If this massacre did not occur in a dumb world we should hear all the horrible tumult of the slaughter-houses of Chicago. But only the ear of the mind can hear the shrieks and lamentations of the eviscerated victims. For myself, I possess this ear, and am full of remorse for having provoked such sufferings.

Now the beetles are rummaging in all directions through the heap of dead and dying, each tugging and tearing at a morsel which he carries off to swallow in peace, away from the inquisitive eyes of his fellows. This mouthful disposed of, another is hastily cut from the body of some victim, and the process is repeated so long as there are bodies left. In a few minutes the procession is reduced to a few shreds of still palpitating flesh.

There were a hundred and fifty caterpillars; the butchers were twenty-five. This amounts to six victims dispatched by each beetle. If the insect had nothing to do but to kill, like the knackers in the meat factories, and if the staff numbered a hundred--a very modest figure as compared with the staff of a lard or bacon factory--then the total number of victims, in a day of ten hours, would be thirty-six thousand. No Chicago "cannery" ever rivalled such a result.

The speed of assassination is the more remarkable when we consider the difficulties of attack. The beetle has no endless chain to seize its victim by one leg, hoist it up, and swing it along to the butcher's knife; it has no sliding plank to hold the victim's head beneath the pole-axe of the knacker; it has to fall upon its prey, overpower it, and avoid its feet and its mandibles. Moreover, the beetle eats its prey on the spot as it kills. What slaughter there would be if the insect confined itself to killing!

What do we learn from the slaughter-houses of Chicago and the fate of the beetle's victims? This: That the man of elevated morality is so far a very rare exception. Under the skin of the civilised being there lurks almost always the ancestor, the savage contemporary of the cave-bear. True humanity does not yet exist; it is growing, little by little, created by the ferment of the centuries and the dictates of conscience; but it progresses towards the highest with heartbreaking slowness.

It was only yesterday that slavery finally disappeared: the basis of the ancient social organism; only yesterday was it realised that man, even though black, is really man and deserves to be treated accordingly.

What formerly was woman? She was what she is to-day in the East: a gentle animal without a soul. The question was long discussed by the learned. The great divine of the seventeenth century, Bossuet himself, regarded woman as the diminutive of man. The proof was in the origin of Eve: she was the superfluous bone, the thirteenth rib which Adam possessed in the beginning. It has at last been admitted that woman possesses a soul like our own, but even superior in tenderness and devotion. She has been allowed to educate herself, which she has done at least as zealously as her coadjutor. But the law, that gloomy cavern which is still the lurking-place of so many barbarities, continues to regard her as an incapable and a minor. The law in turn will finally surrender to the truth.

The abolition of slavery and the education of woman: these are two enormous strides upon the path of moral progress. Our descendants will go farther. They will see, with a lucidity capable of piercing every obstacle, that war is the most hopeless of all absurdities. That our conquerors, victors of battles and destroyers of nations, are detestable scourges; that a clasp of the hand is preferable to a rifle-shot; that the happiest people is not that which possesses the largest battalions, but that which labours in peace and produces abundantly; and that the amenities of existence do not necessitate the existence of frontiers, beyond which we meet with all the annoyances of the custom-house, with its officials who search our pockets and rifle our luggage.

Our descendants will see this and many other marvels which to-day are extravagant dreams. To what ideal height will the process of evolution lead mankind? To no very magnificent height, it is to be feared. We are afflicted by an indelible taint, a kind of original sin, if we may call sin a state of things with which our will has nothing to do. We are made after a certain pattern and we can do nothing to change ourselves. We are marked with the mark of the beast, the taint of the belly, the inexhaustible source of bestiality.

The intestine rules the world. In the midst of our most serious affairs there intrudes the imperious question of bread and butter. So long as there are stomachs to digest--and as yet we are unable to dispense with them--we must find the where-withal to fill them, and the powerful will live by the sufferings of the weak. Life is a void that only death can fill. Hence the endless butchery by which man nourishes himself, no less than beetles and other creatures; hence the perpetual holocausts which make of this earth a knacker's yard, beside which the slaughter-houses of Chicago are as nothing.

But the feasters are legion, and the feast is not abundant in proportion. Those that have not are envious of those that have; the hungry bare their teeth at the satis-fied. Then follows the battle for the right of possession. Man raises armies; to defend his harvests, his granaries, and his cellars, he resorts to warfare. When shall we see the end of it? Alas, and many times alas! As long as there are wolves in the world there must be watch-dogs to defend the flock.

This train of thought has led us far away from our beetles. Let us return to them. What was my motive in provoking the massacre of this peaceful procession of caterpillars who were on the point of self-burial when I gave them over to the butchers? Was it to enjoy the spectacle of a frenzied massacre? By no means; I have always pitied the sufferings of animals, and the smallest life is worthy of respect. To overcome this pity there needed the exigencies of scientific research--exigencies which are often cruel.

In this case the subject of research was the habits of the Carabus auratus, the little vermin-killer of our gardens, who is therefore vulgarly known as the Gar-dener Beetle. How far is this title deserved? What game does the Gardener Beetle hunt? From what vermin does he free our beds and borders? His dealings with the procession of pine-caterpillars promise much. Let us continue our inquiry.

On various occasions about the end of April the gardens afford me the sight of such processions, sometimes longer, sometimes shorter. I capture them and place them in the vivarium. Bloodshed commences the moment the banquet is served. The caterpillars are eviscerated; each by a single beetle, or by several simultane-ously. In less than fifteen minutes the flock is completely exterminated. Nothing remains but a few shapeless fragments, which are carried hither and thither, to be consumed at leisure under the shelter of the wooden board. One well-fed beetle decamps, his booty in his jaws, hoping to finish his feast in peace. He is met by companions who are attracted by the morsel hanging from the mandibles of the fugitive, and audaciously attempt to rob him. First two, then three, they all endeavour to deprive the legitimate owner of his prize. Each seizes the fragment, tugs at it, commences to swallow it without further ado. There is no actual battle; no violent assaults, as in the case of dogs disputing a bone. Their efforts are confined to the attempted theft. If the legitimate owner retains his hold they consume his booty in common, mandibles to mandibles, until the fragment is torn or bitten through, and each retires with his mouthful.

As I found to my cost in bygone experiments, the pine-caterpillar wields a violently corrosive poison, which produces a painful rash upon the hands. It must therefore, one would think, form a somewhat highly seasoned diet. The beetles, however, delight in it. No matter how many flocks I provide them with, they are all consumed. But no one, that I know of, has ever found the Golden Gardener and its larva in the silken cocoons of the Bombyx. I do not expect ever to make such a discovery. These cocoons are inhabited only in winter, when the Gardener is indifferent to food, and lies torpid in the earth. In April, however, when the processions of larvae are seeking a suitable site for burial and metamorphosis, the Gardener should profit largely by its good fortune should it by any chance encounter them.

The furry nature of the victim does not in the least incommode the beetle; but the hairiest of all our caterpillars, the Hedgehog Caterpillar, with its undulating mane, partly red and partly black, does seem to be too much for the beetle. Day after day it wanders about the vivarium in company with the assassins. The latter apparently ignore its presence. From time to time one of them will halt, stroll round the hairy creature, examine it, and try to penetrate the tangled fleece. Immediately repulsed by the long, dense palisade of hairs, he retires without inflicting a wound, and the caterpillar proceeds upon its way with undulating mane, in pride and security.

But this state of things cannot last. In a hungry moment, emboldened moreover by the presence of his fellows, the cowardly creature decides upon a serious attack. There are four of them; they industriously attack the caterpillar, which finally succumbs, assaulted before and behind. It is eviscerated and swallowed as greedily as though it were a defenceless grub.

According to the hazard of discovery, I provision my menagerie with various caterpillars, some smooth and others hairy. All are accepted with the utmost eagerness, so long as they are of average size as compared with the beetles themselves. If too small they are despised, as they would not yield a sufficient mouthful. If they are too large the beetle is unable to handle them. The caterpillars of the Sphinx moth and the Great Peacock moth, for example, would fall an easy prey to the beetle were it not that at the first bite of the assailant the intended victim, by a contortion of its powerful flanks, sends the former flying. After several attacks, all of which end by the beetle being flung back to some considerable distance, the insect regretfully abandons his prey. I have kept two strong and lively caterpillars for a fortnight in the cage of my golden beetles, and nothing more serious occurred. The trick of the suddenly extended posterior was too much for the ferocious mandibles.

The chief utility of the Golden Gardener lies in its extermination of all caterpillars that are not too powerful to attack. It has one limitation, however: it is not a climber. It hunts on the ground; never in the foliage overhead. I have never seen it exploring the twigs of even the smallest of bushes. When caged it pays no attention to the most enticing caterpillars if the latter take refuge in a tuft of thyme, at a few inches above the ground. This is a great pity. If only the beetle could climb how rapidly three or four would rid our cabbages of that grievous pest, the larva of the white cabbage butterfly! Alas! the best have always some failing, some vice.

To exterminate caterpillars: that is the true vocation of the Golden Gardener. It is annoying that it can give us but little or no assistance in ridding us of another plague of the kitchen-garden: the snail. The slime of the snail is offensive to the beetle; it is safe from the latter unless crippled, half crushed, or projecting from the shell. Its relatives, however, do not share this dislike. The horny Procrustes, the great Scarabicus, entirely black and larger than the Carabus, attacks the snail most valiantly, and empties its shell to the bottom, in spite of the desperate secretion of slime. It is a pity that the Procrustes is not more frequently found in our gardens; it would be an excellent gardener's assistant.

CHAPTER IX
THE GOLDEN GARDENER--COURTSHIP

It is generally recognized that the Carabus auratus is an active exterminator of caterpillars; on this account in particular it deserves its title of Gardener Beetle; it is the vigilant policeman of our kitchen-gardens, our flower-beds and herbaceous borders. If my inquiries add nothing to its established reputation in this respect, they will nevertheless, in the following pages, show the insect in a light as yet unsuspected. The ferocious beast of prey, the ogre who devours all creatures that are not too strong for him, is himself killed and eaten: by his fellows, and by many others.

Standing one day in the shadow of the plane-trees that grow before my door, I see a Golden Gardener go by as if on pressing business. The pilgrim is well met; he will go to swell the contents of my vivarium. In capturing him I notice that the extremities of the wing-covers are slightly damaged. Is this the result of a struggle between rivals? There is nothing to tell me. The essential thing is that the insect should not be handicapped by any serious injury. Inspected, and found to be without any serious wound and fit for service, it is introduced into the glass dwelling of its twenty-five future companions.

Next day I look for the new inmate. It is dead. Its comrades have attacked it during the night and have cleaned out its abdomen, insufficiently protected by the damaged wing-covers. The operation has been performed very cleanly, without any dismemberment. Claws, head, corselet, all are correctly in place; the abdomen only has a gaping wound through which its contents have been removed. What remains is a kind of golden shell, formed of the two conjoined elytra. The shell of an oyster emptied of its inmate is not more empty.

This result astonishes me, for I have taken good care that the cage should never be long without food. The snail, the pine-cockchafer, the Praying Mantis, the lob-worm, the caterpillar, and other favourite insects, have all been given in alternation and in sufficient quantities. In devouring a brother whose damaged armour lent itself to any easy attack my beetles had not the excuse of hunger.

Is it their custom to kill the wounded and to eviscerate such of their fellows as suffer damage? Pity is unknown among insects. At the sight of the desperate strug-gles of a crippled fellow-creature none of the same family will cry a halt, none will attempt to come to its aid. Among the carnivorous insects the matter may develop to a tragic termination. With them, the passers-by will often run to the cripple. But do they do so in order to help it? By no means: merely to taste its flesh, and, if they find it agreeable, to perform the most radical cure of its ills by devouring it.

It is possible, therefore, that the Gardener with the injured wing-covers had tempted his fellows by the sight of his imperfectly covered back. They saw in their defenceless comrade a permissible subject for dissection. But do they respect one another when there is no previous wound? At first there was every appearance that their relations were perfectly pacific. During their sanguinary meals there is never a scuffle between the feasters; nothing but mere mouth-to-mouth thefts. There are no quarrels during the long siestas in the shelter of the board. Half buried in the cool earth, my twenty-five subjects slumber and digest their food in peace; they lie sociably near one another, each in his little trench. If I raise the plank they awake and are off, running hither and thither, constantly encountering one another with-out hostilities.

The profoundest peace is reigning, and to all appearances will last for ever, when in the early days of June I find a dead Gardener. Its limbs are intact; it is reduced to the condition of a mere golden husk; like the defenceless beetle I have already spoken of, it is as empty as an oyster-shell. Let us examine the remains. All is intact, save the huge breach in the abdomen. So the insect was sound and unhurt when the others attacked it.

A few days pass, and another Gardener is killed and dealt with as before, with no disorder in the component pieces of its armour. Let us place the dead insect on its belly; it is to all appearances untouched. Place it on its back; it is hollow, and has no trace of flesh left beneath its carapace. A little later, and I find another empty relic; then another, and yet another, until the population of my menagerie is rapidly shrinking. If this insensate massacre continues I shall soon find my cage depopulated.

Are my beetles hoary with age? Do they die a natural death, and do the survivors then clean out the bodies? Or is the population being reduced at the expense of sound and healthy insects? It is not easy to elucidate the matter, since the atrocities are commonly perpetrated in the night. But, finally, with vigilance, on two occasions, I surprise the beetles at their work in the light of day.

Towards the middle of June a female attacks a male before my eyes. The male is recognisable by his slightly smaller size. The operation commences. Raising the ends of the wing-covers, the assailant seizes her victim by the extremity of the abdomen, from the dorsal side. She pulls at him furiously, eagerly munching with her mandibles. The victim, who is in the prime of life, does not defend himself, nor turn upon his assailant. He pulls his hardest in the opposite direction to free himself from those terrible fangs; he advances and recoils as he is overpowered by or overpowers the assassin; and there his resistance ends. The struggle lasts a quarter of an hour. Other beetles, passing by, call a halt, and seem to say "My turn next!" Finally, redoubling his efforts, the male frees himself and flies. If he had not succeeded in escaping the ferocious female would undoubtedly have eviscerated him.

A few days later I witness a similar scene, but this time the tragedy is played to the end. Once more it is a female who seizes a male from behind. With no other protest except his futile efforts to escape, the victim is forced to submit. The skin finally yields; the wound enlarges, and the viscera are removed and devoured by the matron, who empties the carapace, her head buried in the body of her late companion. The legs of the miserable victim tremble, announcing the end. The murderess takes no notice; she continues to rummage as far as she can reach for the narrowing of the thorax. Nothing is left but the closed boat-shaped wing-covers and the fore parts of the body. The empty shell is left lying on the scene of the tragedy.

In this way must have perished the beetles--always males--whose remains I find in the cage from time to time; thus the survivors also will perish. Between the middle of June and the 1st of August the inhabitants of the cage, twenty-five in number at the outset, are reduced to five, all of whom are females. All the males, to the number of twenty, have disappeared, eviscerated and completely emptied. And by whom? Apparently by the females.

That this is the case is attested in the first place by the two assaults of which I was perchance the witness; on two occasions, in broad daylight, I saw the female devouring the male, having opened the abdomen under the wing-covers, or having at least attempted to do so. As for the rest of the massacres, although direct observation was lacking, I had one very valuable piece of evidence. As we have seen, the victim does not retaliate, does not defend himself, but simply tries to escape by pulling himself away.

If it were a matter of an ordinary fight, a conflict such as might arise in the struggle for life, the creature attacked would obviously retaliate, since he is perfectly well able to do so; in an ordinary conflict he would meet force by force, and return bite for bite. His strength would enable him to come well out of a struggle, but the foolish creature allows himself to be devoured without retaliating. It seems as though an invincible repugnance prevents him from offering resistance and in turn devouring the devourer. This tolerance reminds one of the scorpion of Languedoc, which on the termination of the hymeneal rites allows the female to devour him without attempting to employ his weapon, the venomous dagger which would form a formidable defence; it reminds us also of the male of the Praying Mantis, which still embraces the female though reduced to a headless trunk, while the latter devours him by small mouthfuls, with no rebellion or defence on his part. There are other examples of hymeneal rites to which the male offers no resistance.

The males of my menagerie of Gardeners, one and all eviscerated, speak of similar customs. They are the victims of the females when the latter have no further use for them. For four months, from April to August, the insects pair off continually; sometimes tentatively, but usually the mating is effective. The business of mating is all but endless for these fiery spirits.

The Gardener is prompt and businesslike in his affairs of the heart. In the midst of the crowd, with no preliminary courtship, the male throws himself upon the female. The female thus embraced raises her head a trifle as a sign of acquiescence, while the cavalier beats the back of her neck with his antennae. The embrace is brief, and they abruptly separate; after a little refreshment the two parties are ready for other adventures, and yet others, so long as there are males available. After the feast, a brief and primitive wooing; after the wooing, the feast; in such delights the life of the Gardener passes.

The females of my collection were in no proper ratio to the number of aspiring lovers; there were five females to twenty males. No matter; there was no rivalry, no hustling; all went peacefully and sooner or later each was satisfied.

I should have preferred a better proportioned assembly. Chance, not choice, had given me that at my disposal. In the early spring I had collected all the Gardeners I could find under the stones of the neighbourhood, without distinguishing the sexes, for they are not easy to recognise merely by external characteristics. Later on I learned by watching them that a slight excess of size was the distinctive sign of the female. My menagerie, so ill-proportioned in the matter of sex, was therefore the result of chance. I do not suppose this preponderance of males exists in natural conditions. On the other hand, one never sees such numerous groups at liberty, in the shelter of the same stone. The Gardener lives an almost solitary life; it is rarely that one finds two or three beneath the same object of shelter. The gathering in my menagerie was thus exceptional, although it did not lead to confusion. There is plenty of room in the glass cage for excursions to a distance and for all their habitual manoeuvres. Those who wish for solitude can obtain it; those who wish for company need not seek it.

For the rest, captivity cannot lie heavily on them; that is proved by their frequent feasts, their constant mating. They could not thrive better in the open; perhaps not so well, for food is less abundant under natural conditions. In the matter of well-being the prisoners are in a normal condition, favourable to the maintenance of their usual habits.

It is true that encounters of beetle with beetle are more frequent here than in the open. Hence, no doubt, arise more opportunities for the females to persecute the males whom they no longer require; to fall upon them from the rear and eviscerate them. This pursuit of their onetime lovers is aggravated by their confined quarters; but it certainly is not caused thereby, for such customs are not suddenly originated.

The mating season over, the female encountering a male in the open must evidently regard him as fair game, and devour him as the termination of the matrimonial rites. I have turned over many stones, but have never chanced upon this spectacle, but what has occurred in my menagerie is sufficient to convince me. What a world these beetles live in, where the matron devours her mate so soon as her fertility delivers her from the need of him! And how lightly the males must be regarded by custom, to be served in this manner!

Is this practice of post-matrimonial cannibalism a general custom in the insect world? For the moment, I can recollect only three characteristic examples: those of the Praying Mantis, the Golden Gardener, and the scorpion of Languedoc. An analogous yet less brutal practice--for the victim is defunct before he is eaten--is a characteristic of the Locust family. The female of the white-faced Decticus will eagerly devour the body of her dead mate, as will the Green Grasshopper.

To a certain extent this custom is excused by the nature of the insect's diet; the Decticus and the Grasshopper are essentially carnivorous. Encountering a dead body of their own species, a female will devour it, even if it be the body of her latest mate.

But what are we to say in palliation of the vegetarians? At the approach of the breeding season, before the eggs are laid, the Ephippigera turns upon her still living mate, disembowels him, and eats as much of him as her appetite will allow.

The cheerful Cricket shows herself in a new light at this season; she attacks the mate who lately wooed her with such impassioned serenades; she tears his wings, breaks his musical thighs, and even swallows a few mouthfuls of the instrumentalist. It is probable that this deadly aversion of the female for the male at the end of the mating season is fairly common, especially among the carnivorous insects. But what is the object of this atrocious custom? That is a question I shall not fail to answer when circumstances permit.

CHAPTER X
THE FIELD-CRICKET

The breeding of Crickets demands no particular preparations. A little patience is enough--patience, which according to Buffon is genius; but which I, more modestly, will call the superlative virtue of the observer. In April, May, or later we may establish isolated couples in ordinary flower-pots containing a layer of beaten earth. Their diet will consist of a leaf of lettuce renewed from time to time. The pot must be covered with a square of glass to prevent the escape of the inmates.

I have gathered some very curious data from these makeshift appliances, which may be used with and as a substitute for the cages of wire gauze, although the latter are preferable. We shall return to the point presently. For the moment let us watch the process of breeding, taking care that the critical hour does not escape us.

It was during the first week of June that my assiduous visits were at last repaid. I surprised the female motionless, with the oviduct planted vertically in the soil. Heedless of the indiscreet visitor, she remained for a long time stationed at the same point. Finally she withdrew her oviduct, and effaced, though without particular care, the traces of the hole in which her eggs were deposited, rested for a moment, walked away, and repeated the operation; not once, but many times, first here, then there, all over the area at her disposal. Her behaviour was precisely the same as that of the Decticus, except that her movements were more deliberate. At the end of twenty-four hours her eggs were apparently all laid. For greater certainty I waited a couple of days longer.

I then examined the earth in the pot. The eggs, of a straw-yellow, are cylindrical in form, with rounded ends, and measure about one-tenth of an inch in length. They are placed singly in the soil, in a perpendicular position.

I have found them over the whole area of the pot, at a depth of a twelfth of an inch. As closely as the difficulties of the operation will allow, I have estimated the eggs of a single female, upon passing the earth through a sieve, at five or six hundred. Such a family will certainly undergo an energetic pruning before very long.

The egg of the Cricket is a curiosity, a tiny mechanical marvel. After hatching it appears as a sheath of opaque white, open at the summit, where there is a round and very regular aperture, to the edge of which adheres a little valve like a skull-cap which forms the lid. Instead of breaking at random under the thrusts or the cuts of the new-formed larva, it opens of itself along a line of least resistance which occurs expressly for the purpose. The curious process of the actual hatching should be observed.

A fortnight after the egg is laid two large eye-marks, round and of a reddish black, are seen to darken the forward extremity of the egg. Next, a little above these two points, and right at the end of the cylinder, a tiny circular capsule or swelling is seen. This marks the line of rupture, which is now preparing. Presently the translucency of the egg allows us to observe the fine segmentation of the tiny inmate. Now is the moment to redouble our vigilance and to multiply our visits, especially during the earlier part of the day.

Fortune favours the patient, and rewards my assiduity Round the little capsule changes of infinite delicacy have prepared the line of least resistance. The end of the egg, pushed by the head of the inmate, becomes detached, rises, and falls aside like the top of a tiny phial. The Cricket issues like a Jack-in-the-box.

When the Cricket has departed the shell remains distended, smooth, intact, of the purest white, with the circular lid hanging to the mouth of the door of exit. The egg of the bird breaks clumsily under the blows of a wart-like excrescence which is formed expressly upon the beak of the unborn bird; the egg of the Cricket, of a far superior structure, opens like an ivory casket. The pressure of the inmate's head is sufficient to work the hinge.

The moment he is deprived of his white tunic, the young Cricket, pale all over, almost white, begins to struggle against the overlying soil. He strikes it with his mandibles; he sweeps it aside, kicking it backwards and downwards; and being of a powdery quality, which offers no particular resistance, he soon arrives at the surface, and henceforth knows the joys of the sun, and the perils of intercourse with the living; a tiny, feeble creature, little larger than a flea. His colour deepens. In twenty-four hours he assumes a splendid ebony black which rivals that of the adult insect. Of his original pallor he retains only a white girdle which encircles the thorax and reminds one of the leading-string of an infant.

Very much on the alert, he sounds his surroundings with his long vibrating antennae; he toddles and leaps along with a vigour which his future obesity will no longer permit.

This is the age of stomach troubles. What are we to give him to eat? I do not know. I offer him adult diet--the tender leaves of a lettuce. He disdains to bite it; or perhaps his bites escape me, so tiny would they be.

In a few days, what with my ten households, I see myself loaded with family cares. What shall I do with my five or six thousand Crickets, an attractive flock, to be sure, but one I cannot bring up in my ignorance of the treatment required? I will give you liberty, gentle creatures! I will confide you to the sovereign nurse and schoolmistress, Nature!

It is done. Here and there about my orchard, in the most favourable localities, I loose my legions. What a concert I shall have before my door next year if all goes well! But no! There will probably be silence, for the terrible extermination will follow which corresponds with the fertility of the mother. A few couples only may survive: that is the most we can hope.

The first to come to the living feast and the most eager at the slaughter are the little grey lizard and the ant. I am afraid this latter, hateful filibuster that it is, will not leave me a single Cricket in my garden. It falls upon the tiny Crickets, eviscerates them, and devours them with frantic greed.

Satanic creature! And to think that we place it in the front rank of the insect world! The books celebrate its virtues and never tire of its praises; the naturalists hold it in high esteem and add to its reputation daily; so true is it of animals, as of man, that of the various means of living in history the most certain is to do harm to others.

Every one knows the **Bousier** (dung-beetle) and the Necrophorus, those lively murderers; the gnat, the drinker of blood; the wasp, the irascible bully with the poisoned dagger; and the ant, the maleficent creature which in the villages of the South of France saps and imperils the rafters and ceilings of a dwelling with the same energy it brings to the eating of a fig. I need say no more; human history is full of similar examples of the useful misunderstood and undervalued and the calamitous glorified.

What with the ants and other exterminating forces, the massacre was so great that the colonies of Crickets in my orchard, so numerous at the outset, were so far decimated that I could not continue my observations, but had to resort to the outside world for further information.

In August, among the detritus of decaying leaves, in little oases whose turf is not burned by the sun, I find the young Cricket has already grown to a considerable size; he is all black, like the adult, without a vestige of the white cincture of the early days. He has no domicile. The shelter of a dead leaf, the cover afforded by a flat stone is sufficient; he is a nomad, and careless where he takes his repose.

Not until the end of October, when the first frosts are at hand, does the work of burrowing commence. The operation is very simple, as far as I can tell from what I have learned from the insect in captivity. The burrow is never made at a bare or conspicuous point; it is always commenced under the shelter of a faded leaf of lettuce, the remains of the food provided. This takes the place of the curtain of grass so necessary to preserve the mysterious privacy of the establishment.

The little miner scratches with his fore-claws, but also makes use of the pincers of his mandibles in order to remove pieces of grit or gravel of any size. I see him stamping with his powerful hinder limbs, which are provided with a double row of spines; I see him raking and sweeping backwards the excavated material, and spreading it out in an inclined plane. This is his whole method.

At first the work goes forward merrily. The excavator disappears under the easily excavated soil of his prison after two hours' labour. At intervals he returns to the orifice, always tail first, and always raking and sweeping. If fatigue overcomes him he rests on the threshold of his burrow, his head projecting outwards, his antennae gently vibrating. Presently he re-enters his tunnel and sets to work again with his pincers and rakes. Presently his periods of repose grow longer and tire my patience.

The most important part of the work is now completed. Once the burrow has attained a depth of a couple of inches, it forms a sufficient shelter for the needs of the moment. The rest will be the work of time; a labour resumed at will, for a short time daily. The burrow will be made deeper and wider as the growth of the inmate and the inclemency of the season demand. Even in winter, if the weather is mild, and the sun smiles upon the threshold of his dwelling, one may sometimes surprise the Cricket thrusting out small quantities of loosened earth, a sign of enlargement and of further burrowing. In the midst of the joys of spring the cares of the house still continue; it is constantly restored and perfected until the death of the occupant.

April comes to an end, and the song of the Cricket commences. At first we hear only timid and occasional solos; but very soon there is a general symphony, when every scrap of turf has its performer. I am inclined to place the Cricket at the head of the choristers of spring. In the waste lands of Provence, when the thyme and the lavender are in flower, the Cricket mingles his note with that of the crested lark, which ascends like a lyrical firework, its throat swelling with music, to its invisible station in the clouds, whence it pours its liquid arias upon the plain below. From the ground the chorus of the Crickets replies. It is monotonous and artless, yet how well it harmonises, in its very simplicity, with the rustic gaiety of a world renewed! It is the hosanna of the awakening, the alleluia of the germinating seed and the sprouting blade. To which of the two performers should the palm be given? I should award it to the Cricket; he triumphs by force of numbers and his never-ceasing note. The lark hushes her song, that the blue-grey fields of lavender, swinging their aromatic censers before the sun, may hear the Cricket alone at his humble, solemn celebration.

But here the anatomist intervenes, roughly demanding of the Cricket: "Show me your instrument, the source of your music!" Like all things of real value, it is very simple; it is based on the same principle as that of the locusts; there is the toothed fiddlestick and the vibrating tympanum.

The right wing-cover overlaps the left and almost completely covers it, except for the sudden fold which encases the insect's flank. This arrangement is the reverse of that exhibited by the green grasshopper, the Decticus, the Ephippigera, and their relations. The Cricket is right-handed, the others left-handed. The two wing-covers have the same structure. To know one is to know the other. Let us examine that on the right hand.

It is almost flat on the back, but suddenly folds over at the side, the turn being almost at right angles. This lateral fold encloses the flank of the abdomen and is covered with fine oblique and parallel nervures. The powerful nervures of the dorsal portion of the wing-cover are of the deepest black, and their general effect is that of a complicated design, not unlike a tangle of Arabic caligraphy.

Seen by transmitted light the wing-cover is of a very pale reddish colour, excepting two large adjacent spaces, one of which, the larger and anterior, is triangular in shape, while the other, the smaller and posterior, is oval. Each space is surrounded by a strong nervure and goffered by slight wrinkles or depressions. These two spaces represent the mirror of the locust tribe; they constitute the sonorous area. The substance of the wing-cover is finer here than elsewhere, and shows traces of iridescent though somewhat smoky colour.

These are parts of an admirable instrument, greatly superior to that of the Decticus. The five hundred prisms of the bow biting upon the ridges of the wing-cover opposed to it set all four tympanums vibrating at once; the lower pair by direct friction, the upper pair by the vibration of the wing-cover itself. What a powerful sound results! The Decticus, endowed with only one indifferent "mirror," can be heard only at a few paces; the Cricket, the possessor of four vibratory areas, can be heard at a hundred yards.

The Cricket rivals the Cigale in loudness, but his note has not the displeasing, raucous quality of the latter. Better still: he has the gift of expression, for he can sing loud or soft. The wing-covers, as we have seen, are prolonged in a deep fold over each flank. These folds are the dampers, which, as they are pressed downwards or slightly raised, modify the intensity of the sound, and according to the extent of their contact with the soft abdomen now muffle the song to a *mezza voce* and now let it sound *fortissimo*.

Peace reigns in the cage until the warlike instinct of the mating period breaks out. These duels between rivals are frequent and lively, but not very serious. The two rivals rise up against one another, biting at one another's heads--these solid, fang-proof helmets--roll each other over, pick themselves up, and separate. The vanquished Cricket scuttles off as fast as he can; the victor insults him by a couple of triumphant and boastful chirps; then, moderating his tone, he tacks and veers about the desired one.

The lover proceeds to make himself smart. Hooking one of his antennae towards him with one of his free claws, he takes it between his mandibles in order to curl it and moisten it with saliva. With his long hind legs, spurred and laced with red, he stamps with impatience and kicks out at nothing. Emotion renders him silent. His wing-covers are nevertheless in rapid motion, but are no longer sounding, or at most emit but an unrhythmical rubbing sound.

Presumptuous declaration! The female Cricket does not run to hide herself in the folds of her lettuce leaves; but she lifts the curtain a little, and looks out, and wishes to be seen:--

Et fugit ad salices, et se cupit ante videri.

She flies towards the brake, but hopes first to be perceived, said the poet of the delightful eclogue, two thousand years ago. Sacred provocations of lovers, are they not in all ages the same?

CHAPTER XI
THE ITALIAN CRICKET

My house shelters no specimens of the domestic Cricket, the guest of bakeries and rustic hearths. But although in my village the chinks under the hearthstones are mute, the nights of summer are musical with a singer little known in the North. The sunny hours of spring have their singer, the Field-Cricket of which I have written; while in the summer, during the stillness of the night, we hear the note of the Italian Cricket, the *OEcanthus pellucens*, Scop. One diurnal and one nocturnal, between them they share the kindly half of the year. When the Field-Cricket ceases to sing it is not long before the other begins its serenade.

The Italian Cricket has not the black costume and heavy shape characteristic of the family. It is, on the contrary, a slender, weakly creature; its colour very pale, indeed almost white, as is natural in view of its nocturnal habits. In handling it one is afraid of crushing it between the fingers. It lives an aerial existence; on shrubs and bushes of all kinds, on tall herbage and grasses, and rarely descends to the earth. Its song, the pleasant voice of the calm, hot evenings from July to October, commences at sunset and continues for the greater part of the night.

This song is familiar to all Provencals; for the least patch of thicket or tuft of grasses has its group of instrumentalists. It resounds even in the granaries, into which the insect strays, attracted thither by the fodder. But no one, so mysterious are the manners of the pallid Cricket, knows exactly what is the source of the serenade, which is often, though quite erroneously, attributed to the common field-cricket, which at this period is silent and as yet quite young.

The song consists of a *Gri-i-i, Gri-i-i*, a slow, gentle note, rendered more expressive by a slight tremor. Hearing it, one divines the extreme tenuity and the amplitude of the vibrating membranes. If the insect is not in any way disturbed as it sits in the low foliage, the note does not vary, but at the least noise the performer becomes a ventriloquist. First of all you hear it there, close by, in front of you, and the next moment you hear it over there, twenty yards away; the double note decreased in volume by the distance.

You go forward. Nothing is there. The sound proceeds again from its original point. But no--it is not there; it is to the left now--unless it is to the right--or behind.... Complete confusion! It is impossible to detect, by means of the ear, the direction from which the chirp really comes. Much patience and many precautions will be required before you can capture the insect by the light of the lantern. A few specimens caught under these conditions and placed in a cage have taught me the little I know concerning the musician who so perfectly deceives our ears.

The wing-covers are both formed of a dry, broad membrane, diaphanous and as fine as the white skin on the outside of an onion, which is capable of vibrating over its whole area. Their shape is that of the segment of a circle, cut away at the upper end. This segment is bent at a right angle along a strong longitudinal nervure, and descends on the outer side in a flap which encloses the insect's flank when in the attitude of repose.

The right wing-cover overlaps the left. Its inner edge carries, on the under side, near the base, a callosity from which five radiating nervures proceed; two of them upwards and two downwards, while the fifth runs approximately at right angles to these. This last nervure, which is of a slightly reddish hue, is the fundamental element of the musical device; it is, in short, the bow, the fiddlestick, as is proved by the fine notches which run across it. The rest of the wing-cover shows a few more nervures of less importance, which hold the membrane stretched tight, but do not form part of the friction apparatus.

The left or lower wing-cover is of similar structure, with the difference that the bow, the callosity, and the nervures occupy the upper face. It will be found that the two bows--that is, the toothed or indented nervures--cross one another obliquely.

When the note has its full volume, the wing-covers are well raised above the body like a wide gauzy sail, only touching along the internal edges. The two bows, the toothed nervures, engage obliquely one with the other, and their mutual friction causes the sonorous vibration of the two stretched membranes.

The sound can be modified accordingly as the strokes of each bow bear upon the callosity, which is itself serrated or wrinkled, or on one of the four smooth radiating nervures. Thus in part are explained the illusions produced by a sound which seems to come first from one point, then from another, when the timid insect is alarmed.

The production of loud or soft resounding or muffled notes, which gives the illusion of distance, the principal element in the art of the ventriloquist, has another and easily discovered source. To produce the loud, open sounds the wing-covers are fully lifted; to produce the muted, muffled notes they are lowered. When lowered their outer edges press more or less lightly on the soft flanks of the insect, thus diminishing the vibratory area and damping the sound.

The gentle touch of a finger-tip muffles the sharp, loud ringing of a glass tumbler or "musical-glass" and changes it into a veiled, indefinite sound which seems to come from a distance. The White Cricket knows this secret of acoustics. It misleads those that seek it by pressing the edge of its vibrating membranes to the soft flesh of its abdomen. Our musical instruments have their dampers; that of the *OEcanthus pellucens* rivals and surpasses them in simplicity of means and perfection of results.

The Field-Cricket and its relatives also vary the volume of their song by raising or lowering the elytra so as to enclose the abdomen in a varying degree, but none of them can obtain by this method results so deceptive as those produced by the Italian Cricket.

To this illusion of distance, which is a source of perpetually renewed surprise, evoked by the slightest sound of our footsteps, we must add the purity of the sound, and its soft tremolo. I know of no insect voice more gracious, more limpid, in the profound peace of the nights of August. How many times, *per amica silentia lunae*, have I lain upon the ground, in the shelter of a clump of rosemary, to listen to the delicious concert!

The nocturnal Cricket sings continually in the gardens. Each tuft of the red-flowered cistus has its band of musicians, and each bush of fragrant lavender. The shrubs and the terebinth-trees contain their orchestras. With its clear, sweet voice, all this tiny world is questioning, replying, from bush to bush, from tree to tree; or rather, indifferent to the songs of others, each little being is singing his joys to himself alone.

Above my head the constellation of Cygnus stretches its great cross along the Milky Way; below, all around me, palpitates the insect symphony. The atom telling of its joys makes me forget the spectacle of the stars. We know nothing of these celestial eyes which gaze upon us, cold and calm, with scintillations like the blinking of eyelids.

Science tells us of their distance, their speeds, their masses, their volumes; it burdens us with stupendous numbers and stupefies us with immensities; but it does not succeed in moving us. And why? Because it lacks the great secret: the secret of life. What is there, up there? What do these suns warm? Worlds analogous to ours, says reason; planets on which life is evolving in an endless variety of forms. A superb conception of the universe, but after all a pure conception, not based upon patent facts and infallible testimony at the disposal of one and all. The probable, even the extremely probable, is not the obvious, the evident, which forces itself irresistibly and leaves no room for doubt.

But in your company, O my Crickets, I feel the thrill of life, the soul of our native lump of earth; and for this reason, as I lean against the hedge of rosemary, I bestow only an absent glance upon the constellation of Cygnus, but give all my attention to your serenade. A little animated slime, capable of pleasure and pain, surpasses in interest the universe of dead matter.

CHAPTER XII
THE SISYPHUS BEETLE.--THE INSTINCT
OF PATERNITY

The duties of paternity are seldom imposed on any but the higher animals. They are most notable in the bird; and the furry peoples acquit themselves honourably. Lower in the scale we find in the father a general indifference as to the fate of the family. Very few insects form exceptions to this rule. Although all are imbued with a mating instinct that is almost frenzied, nearly all, when the passion of the moment is appeased, terminate then and there their domestic relations, and withdraw, indifferent to the brood, which has to look after itself as best it may.

This paternal coldness, which would be odious in the higher walks of animal life, where the weakness of the young demands prolonged assistance, has in the insect world the excuse that the new-born young are comparatively robust, and are able, without help, to fill their mouths and stomachs, provided they find themselves in propitious surroundings. All that the prosperity of the race demands of the Pierides, or Cabbage Butterflies, is that they should deposit their eggs on the leaves of the cabbage; what purpose would be served by the instincts of a father? The botanical instinct of the mother needs no assistance. At the period of laying the father would be in the way. Let him pursue his flirtations elsewhere; the laying of eggs is a serious business.

In the case of the majority of insects the process of education is unknown, or summary in the extreme. The insect has only to select a grazing-ground upon which its family will establish itself the moment it is hatched; or a site which will allow the young to find their proper sustenance for themselves. There is no need of a father in these various cases. After mating, the discarded male, who is henceforth useless, drags out a lingering existence of a few days, and finally perishes without having given the slightest assistance in the work of installing his offspring.

But matters are not everywhere so primitive as this. There are tribes in which an inheritance is prepared for the family which will assure it both of food and of shelter in advance. The Hymenoptera in particular are past-masters in the provision of cellars, jars, and other utensils in which the honey-paste destined for the young is stored; they are perfect in the art of excavating storehouses of food for their grubs.

This stupendous labour of construction and provisioning, this labour that absorbs the insect's whole life, is the work of the mother only, who wears herself out at her task. The father, intoxicated with sunlight, lies idle on the threshold of the workshop, watching the heroic female at her work, and regards himself as excused from all labour when he has plagued his neighbours a little.

Does he never perform useful work? Why does he not follow the example of the swallows, each of whom brings a fair share of the straw and mortar for the building of the nest and the midges for the young brood? No, he does nothing; perhaps alleging the excuse of his relative weakness. But this is a poor excuse; for to cut out little circles from a leaf, to rake a little cotton from a downy plant, or to gather a little mortar from a muddy spot, would hardly be a task beyond his powers. He might very well collaborate, at least as labourer; he could at least gather together the materials for the more intelligent mother to place in position. The true motive of his idleness is ineptitude.

It is a curious thing that the Hymenoptera, the most skilful of all industrial insects, know nothing of paternal labour. The male of the genus, in whom we should expect the requirements of the young to develop the highest aptitudes, is as useless as a butterfly, whose family costs so little to establish. The actual distribution of instinct upsets our most reasonable previsions.

It upsets our expectations so completely that we are surprised to find in the dung-beetle the noble prerogative which is lacking in the bee tribe. The mates of several species of dung-beetle keep house together and know the worth of mutual labour. Consider the male and female Geotrupes, which prepare together the patrimony of their larvae; in their case the father assists his companion with the pressure of his robust body in the manufacture of their balls of compressed nutriment. These domestic habits are astonishing amidst the general isolation.

To this example, hitherto unique, my continual researches in this direction permit me to-day to add three others which are fully as interesting. All three are members of the corporation of dung-beetles. I will relate their habits, but briefly, as in many respects their history is the same as that of the Sacred Scarabaeus, the Spanish Copris, and others.

The first example is the Sisyphus beetle (Sisyphus Schaefferi, Lin.), the smallest and most industrious of our pill-makers. It has no equal in lively agility, grotesque somersaults, and sudden tumbles down the impossible paths or over the impracticable obstacles to which its obstinacy is perpetually leading it. In allusion to these frantic gymnastics Latreille has given the insect the name of Sisyphus, after the celebrated inmate of the classic Hades. This unhappy spirit underwent terrible exertions in his efforts to heave to the top of a mountain an enormous rock, which always escaped him at the moment of attaining the summit, and rolled back to the foot of the slope. Begin again, poor Sisyphus, begin again, begin again always! Your torments will never cease until the rock is firmly placed upon the summit of the mountain.

I like this myth. It is, in a way, the history of many of us; not odious scoundrels worthy of eternal torments, but worthy and laborious folk, useful to their neighbours. One crime alone is theirs to expiate: the crime of poverty. Half a century or more ago, for my own part, I left many blood-stained tatters on the crags of the inhospitable mountain; I sweated, strained every nerve, exhausted my veins, spent without reckoning my reserves of energy, in order to carry upward and lodge in a place of security that crushing burden, my daily bread; and hardly was the load balanced but it once more slipped downwards, fell, and was engulfed. Begin again, poor Sisyphus; begin again, until your burden, falling for the last time, shall crush your head and set you free at length.

The Sisyphus of the naturalists knows nothing of these tribulations. Agile and lively, careless of slope or precipice, he trundles his load, which is sometimes food for himself, sometimes for his offspring. He is very rare hereabouts; I should never have succeeded in obtaining a sufficient number of specimens for my purpose but for an assistant whom I may opportunely present to the reader, for he will be mentioned again in these recitals.

This is my son, little Paul, aged seven. An assiduous companion of the chase, he knows better than any one of his age the secrets of the Cigale, the Cricket, and especially of the dung-beetle, his great delight. At a distance of twenty yards his clear sight distinguishes the refuse-tip of a beetle's burrow from a chance lump of earth; his fine ear will catch the chirping of a grasshopper inaudible to me. He lends me his sight and hearing, and I in return make him free of my thoughts, which he welcomes attentively, raising his wide blue eyes questioningly to mine.

What an adorable thing is the first blossoming of the intellect! Best of all ages is that when the candid curiosity awakens and commences to acquire knowledge of every kind. Little Paul has his own insectorium, in which the Scarabaeus makes his balls; his garden, the size of a handkerchief, in which he grows haricot beans, which are often dug up to see if the little roots are growing longer; his plantation, containing four oak-trees an inch in height, to which the acorns still adhere. These serve as diversions after the arid study of grammar, which goes forward none the worse on that account.

What beautiful and useful knowledge the teaching of natural history might put into childish heads, if only science would consider the very young; if our barracks of universities would only combine the lifeless study of books with the living study of the fields; if only the red tape of the curriculum, so dear to bureaucrats, would not strangle all willing initiative. Little Paul and I will study as much as possible in the open country, among the rosemary bushes and arbutus. There we shall gain vigour of body and of mind; we shall find the true and the beautiful better than in school-books.

To-day the blackboard has a rest; it is a holiday. We rise early, in view of the intended expedition; so early that we must set out fasting. But no matter; when we are hungry we shall rest in the shade, and you will find in my knapsack the usual viaticum--apples and a crust of bread. The month of May is near; the Sisyphus should have appeared. Now we must explore at the foot of the mountain, the scanty pastures through which the herds have passed; we must break with our fingers, one by one, the cakes of sheep-dung dried by the sun, but still retaining a spot of moisture in the centre. There we shall find Sisyphus, cowering and waiting until the evening for fresher pasturage.

Possessed of this secret, which I learned from previous fortuitous discoveries, little Paul immediately becomes a master in the art of dislodging the beetle. He shows such zeal, has such an instinct for likely hiding-places, that after a brief search I am rich beyond my ambitions. Behold me the owner of six couples of Sisyphus beetles: an unheard-of number, which I had never hoped to obtain.

For their maintenance a wire-gauze cover suffices, with a bed of sand and diet to their taste. They are very small, scarcely larger than a cherry-stone. Their shape is extremely curious. The body is dumpy, tapering to an acorn-shaped posterior; the legs are very long, resembling those of the spider when outspread; the hinder legs are disproportionately long and curved, being thus excellently adapted to enlace and press the little pilule of dung.

Mating takes place towards the beginning of May, on the surface of the soil, among the remains of the sheep-dung on which the beetles have been feeding. Soon the moment for establishing the family arrives. With equal zeal the two partners take part in the kneading, transport, and baking of the food for their offspring. With the file-like forelegs a morsel of convenient size is shaped from the piece of dung placed in the cage. Father and mother manipulate the piece together, striking it blows with their claws, compressing it, and shaping it into a ball about the size of a big pea.

As in the case of the **Scarabaeus sacer**, the exact spherical form is produced without the mechanical device of rolling the ball. Before it is moved, even before it is cut loose from its point of support, the fragment is modelled into the shape of a sphere. The beetle as geometer is aware of the form best adapted to the long preservation of preserved foods.

The ball is soon ready. It must now be forced to acquire, by means of a vigorous rolling, the crust which will protect the interior from a too rapid evaporation. The mother, recognisable by her slightly robuster body, takes the place of honour in front. Her long hinder legs on the soil, her forelegs on the ball, she drags it towards her as she walks backwards. The father pushes behind, moving tail first, his head held low. This is exactly the method of the Scarabaeus beetles, which also work in couples, though for another object. The Sisyphus beetles harness themselves to provide an inheritance for their larvae; the larger insects are concerned in obtaining the material for a banquet which the two chance-met partners will consume underground.

The couple set off, with no definite goal ahead, across the irregularities of the soil, which cannot be avoided by a leader who hauls backwards. But even if the Sisyphus saw the obstacles she would not try to evade them: witness her obstinate endeavour to drag her load up the wire gauze of her cage!

A hopeless undertaking! Fixing her hinder claws in the meshes of the wire gauze the mother drags her burden towards her; then, enlacing it with her legs, she holds it suspended. The father, finding no purchase for his legs, clutches the ball, grows on to it, so to speak, thus adding his weight to that of the burden, and awaits events. The effort is too great to last. Ball and beetle fall together. The mother, from above, gazes a moment in surprise, and suddenly lets herself fall, only to re-embrace the ball and recommence her impracticable efforts to scale the wall. After many tumbles the attempt is at last abandoned.

Even on level ground the task is not without its difficulties. At every moment the load swerves on the summit of a pebble, a fragment of gravel; the team are over-turned, and lie on their backs, kicking their legs in the air. This is a mere nothing. They pick themselves up and resume their positions, always quick and lively. The accidents which so often throw them on their backs seem to cause them no concern; one would even think they were invited. The pilule has to be matured, given a proper consistency. In these conditions falls, shocks, blows, and jolts might well enter into the programme. This mad trundling lasts for hours and hours.

Finally, the mother, considering that the matter has been brought to a satisfactory conclusion, departs in search of a favourable place for storage. The father, crouched upon the treasure, waits. If the absence of his companion is prolonged he amuses himself by rapidly whirling the pill between his hind legs, which are raised in the air. He juggles with the precious burden; he tests its perfections between his curved legs, calliper-wise. Seeing him frisking in this joyful occupation, who can doubt that he experiences all the satisfactions of a father assured of the future of his family? It is I, he seems to say, it is I who have made this loaf, so beautifully round; it is I who have made the hard crust to preserve the soft dough; it is I who have baked it for my sons! And he raises on high, in the sight of all, this magnificent testimonial of his labours.

But now the mother has chosen the site. A shallow pit is made, the mere commencement of the projected burrow. The ball is pushed and pulled until it is close at hand. The father, a vigilant watchman, still retains his hold, while the mother digs with claws and head. Soon the pit is deep enough to receive the ball; she cannot dispense with the close contact of the sacred object; she must feel it bobbing behind her, against her back, safe from all parasites and robbers, before she can decide to burrow further. She fears what might happen to the precious loaf if it were abandoned at the threshold of the burrow until the completion of the dwelling. There is no lack of midges and tiny dung-beetles--Aphodiinae--which might take possession of it. It is only prudent to be distrustful.

So the ball is introduced into the pit, half in and half out of the mouth of the burrow. The mother, below, clasps and pulls; the father, above, moderates the jolts and prevents it from rolling. All goes well. Digging is resumed, and the descent continues, always with the same prudence; one beetle dragging the load, the other regulating its descent and clearing away all rubbish that might hinder the operation. A few more efforts, and the ball disappears underground with the two miners. What follows will be, for a time at least, only a repetition of what we have seen. Let us wait half a day or so.

If our vigilance is not relaxed we shall see the father regain the surface alone, and crouch in the sand near the mouth of the burrow. Retained by duties in the performance of which her companion can be of no assistance, the mother habitually delays her reappearance until the following day. When she finally emerges the father wakes up, leaves his hiding place, and rejoins her. The reunited couple return to their pasturage, refresh themselves, and then cut out another ball of dung. As before, both share the work; the hewing and shaping, the transport, and the burial in ensilage.

This conjugal fidelity is delightful; but is it really the rule? I should not dare to affirm that it is. There must be flighty individuals who, in the confusion under a large cake of droppings, forget the fair confectioners for whom they have worked as journeymen, and devote themselves to the services of others, encountered by chance; there must be temporary unions, and divorces after the burial of a single pellet. No matter: the little I myself have seen gives me a high opinion of the domestic morals of the Sisyphus.

Let us consider these domestic habits a little further before coming to the contents of the burrow. The father works fully as hard as the mother at the extraction and modelling of the pellet which is destined to be the inheritance of a larva; he shares in the work of transport, even if he plays a secondary part; he watches over the pellet when the mother is absent, seeking for a suitable site for the excavation of the cellar; he helps in the work of digging; he carries away the rubbish from the burrow; finally, to crown all these qualities, he is in a great measure faithful to his spouse.

The Scarabaeus exhibits some of these characteristics. He also assists his spouse in the preparation of pellets of dung; he also assists her to transport the pellets, the pair facing each other and the female going backwards. But as I have stated already, the motive of this mutual service is selfish; the two partners labour only for their own good. The feast is for themselves alone. In the labours that concern the family the female Scarabaeus receives no assistance. Alone she moulds her sphere, extracts it from the lump and rolls it backwards, with her back to her task, in the position adopted by the male Sisyphus; alone she excavates her burrow, and alone she buries the fruit of her labour. Oblivious of the gravid mother and the future brood, the male gives her no assistance in her exhausting task. How different to the little pellet-maker, the Sisyphus!

It is now time to visit the burrow. At no very great depth we find a narrow chamber, just large enough for the mother to move around at her work. Its very exiguity proves that the male cannot remain underground; so soon as the chamber is ready he must retire in order to leave the female room to move. We have, in fact, seen that he returns to the surface long before the female.

The contents of the cellar consist of a single pellet, a masterpiece of plastic art. It is a miniature reproduction of the pear-shaped ball of the Scarabaeus, a reproduction whose very smallness gives an added value to the polish of the surface and the beauty of its curves. Its larger diameter varies from half to three-quarters of an inch. It is the most elegant product of the dung-beetle's art.

But this perfection is of brief duration. Very soon the little "pear" becomes covered with gnarled excrescences, black and twisted, which disfigure it like so many warts. Part of the surface, which is otherwise intact, disappears under a shapeless mass. The origin of these knotted excrescences completely deceived me at first. I suspected some cryptogamic vegetation, some **Spheriaecaea**, for example, recognisable by its black, knotted, incrusted growth. It was the larva that showed me my mistake.

The larva is a maggot curved like a hook, carrying on its back an ample pouch or hunch, forming part of its alimentary canal. The reserve of excreta in this hunch enables it to seal accidental perforations of the shell of its lodging with an instantaneous jet of mortar. These sudden emissions, like little worm-casts, are also practised by the Scarabaeus, but the latter rarely makes use of them.

The larvae of the various dung-beetles utilise their alimentary residues in rough-casting their houses, which by their dimensions lend themselves to this method of disposal, while evading the necessity of opening temporary windows by which the ordure can be expelled. Whether for lack of sufficient room, or for other reasons which escape me, the larva of the Sisyphus, having employed a certain amount in the smoothing of the interior, ejects the rest of its digestive products from its dwelling.

Let us examine one of these "pears" when the inmate is already partly grown. Sooner or later we shall see a spot of moisture appear at some point on the surface; the wall softens, becomes thinner, and then, through the softened shell, a jet of dark green excreta rises and falls back upon itself in corkscrew convolutions. One excrescence the more has been formed; as it dries it becomes black.

What has occurred? The larva has opened a temporary breach in the wall of its shell; and through this orifice, in which a slight thickness of the outer glaze still remains, it has expelled the excess of mortar which it could not employ within. This practice of forming oubliettes in the shell of its prison does not endanger the grub, as they are immediately closed, and hermetically sealed by the base of the jet, which is compressed as by a stroke of a trowel. The stopper is so quickly put in place that the contents remain moist in spite of the frequent breaches made in the shell of the "pear." There is no danger of an influx of the dry outer air.

The Sisyphus seems to be aware of the peril which later on, in the dog-days, will threaten its "pear," small as it is, and so near the surface of the ground. It is extremely precocious. It labours in April and May when the air is mild. In the first fortnight of July, before the terrible dog-days have arrived, the members of its family break their shells and set forth in search of the heap of droppings which will furnish them with food and lodging during the fierce days of summer. Then come the short but pleasant days of autumn, the retreat underground and the winter torpor, the awakening of spring, and finally the cycle is closed by the festival of pellet-making.

One word more as to the fertility of the Sisyphus. My six couples under the wire-gauze cover furnished me with fifty-seven inhabited pellets. This gives an average of more than nine to each couple; a figure which the *Scarabaeus sacer* is far from attaining. To what should we attribute this superior fertility? I can only see one cause: the fact that the male works as valiantly as the female. Family cares too great for the strength of one are not too heavy when there are two to support them.

CHAPTER XIII
A BEE-HUNTER: THE *PHILANTHUS AVIPORUS*

To encounter among the Hymenoptera, those ardent lovers of flowers, a species which goes a-hunting on its own account is, to say the least of it, astonishing. That the larder of the larvae should be provisioned with captured prey is natural enough; but that the provider, whose diet is honey, should itself devour its captives is a fact both unexpected and difficult to comprehend. We are surprised that a drinker of nectar should become a drinker of blood. But our surprise abates if we consider the matter closely. The double diet is more apparent than real; the stomach which fills itself with the nectar of flowers does not gorge itself with flesh. When she perforates the rump of her victim the Odynerus does not touch the flesh, which is a diet absolutely contrary to her tastes; she confines herself to drinking the defensive liquid which the grub distils at the end of its intestine. For her this liquid is doubtless a beverage of delicious flavour, with which she relieves from time to time her staple diet of the honey distilled by flowers, some highly spiced condiment, appetiser or aperient, or perhaps--who knows?--a substitute for honey. Although the qualities of the liquid escape me, I see at least that Odynerus cares nothing for the rest. Once the pouch is emptied the larva is abandoned as useless offal, a certain sign of non-carnivorous appetites. Under these conditions the persecutor of Chrysomela can no longer be regarded as guilty of an unnatural double dietary.

We may even wonder whether other species also are not apt to draw some direct profit from the hunting imposed upon them by the needs of the family. The procedure of Odynerus in opening the anal pouch is so far removed from the usual that we should not anticipate many imitators; it is a secondary detail, and impracticable with game of a different kind. But there may well be a certain amount of

variety in the means of direct utilisation. Why, for example, when the victim which
has just been paralysed or rendered insensible by stinging contains in the stomach
a delicious meal, semi-liquid or liquid in consistency, should the hunter scruple to
rob the half-living body and force it to disgorge without injuring the quality of its
flesh? There may well be robbers of the moribund, attracted not by their flesh but
by the appetising contents of their stomachs.

As a matter of fact there are such, and they are numerous. In the first rank we
may cite that hunter of the domestic bee, ***Philanthus aviporus*** (Latreille). For a
long time I suspected Philanthus of committing such acts of brigandage for her own
benefit, having many times surprised her gluttonously licking the honey-smeared
mouth of the bee; I suspected that her hunting of the bee was not undertaken entirely for the benefit of her larvae. The suspicion was worth experimental confirmation. At the time I was interested in another question also: I wanted to study,
absolutely at leisure, the methods by which the various predatory species dealt with
their victims. In the case of Philanthus I made use of the improvised cage already
described; and Philanthus it was who furnished me with my first data on the subject.
She responded to my hopes with such energy that I thought myself in possession of
an unequalled method of observation, by means of which I could witness again and
again, to satiety even, incidents of a kind so difficult to surprise in a state of nature.
Alas! the early days of my acquaintance with Philanthus promised me more than
the future had in store for me! Not to anticipate, however, let us place under the
bell-glass the hunter and the game. I recommend the experiment to whomsoever
would witness the perfection with which the predatory Hymenoptera use their
stings. The result is not in doubt and the waiting is short; the moment the prey is
perceived in an attitude favourable to her designs, the bandit rushes at it, and all is
over. In detail, the tragedy develops as follows:

I place under a bell-glass a Philanthus and two or three domestic bees. The
prisoners climb the glass walls, on the more strongly lighted side; they ascend, descend, and seek to escape; the polished, vertical surface is for them quite easy to
walk upon. They presently quiet down, and the brigand begins to notice her surroundings. The antennae point forward, seeking information; the hinder legs are
drawn up with a slight trembling, as of greed and rapacity, in the thighs; the head
turns to the right and the left, and follows the evolutions of the bees against the

glass. The posture of the scoundrelly insect is strikingly expressive; one reads in it the brutal desires of a creature in ambush, the cunning patience that postpones attack. The choice is made, and Philanthus throws herself upon her victim.

Turn by turn tumbled and tumbling, the two insects roll over and over. But the struggle soon quiets down, and the assassin commences to plunder her prize. I have seen her adopt two methods. In the first, more usual than the other, the bee is lying on the ground, upon its back, and Philanthus, mouth to mouth and abdomen to abdomen, clasps it with her six legs, while she seizes its neck in her mandibles. The abdomen is then curved forward and gropes for a moment for the desired spot in the upper part of the thorax, which it finally reaches. The sting plunges into the victim, remains in the wound for a moment, and all is over. Without loosing the victim, which is still tightly clasped, the murderer restores her abdomen to the normal position and holds it pressed against that of the bee.

By the second method Philanthus operates standing upright. Resting on the hinder feet and the extremity of the folded wings, she rises proudly to a vertical position, holding the bee facing her by her four anterior claws. In order to get the bee into the proper position for the final stroke, she swings the poor creature round and back again with the careless roughness of a child dandling a doll. Her pose is magnificent, solidly based upon her sustaining tripod, the two posterior thighs and the end of the wings, she flexes the abdomen forwards and upwards, and, as before, stings the bee in the upper part of the thorax. The originality of her pose at the moment of striking surpasses anything I have ever witnessed.

The love of knowledge in matters of natural history is not without its cruelties. To make absolutely certain of the point attained by the sting, and to inform myself completely concerning this horrible talent for murder, I have provoked I dare not confess how many assassinations in captivity. Without a single exception, the bee has always been stung in the throat. In the preparations for the final blow the extremity of the abdomen may of course touch here and there, at different points of the thorax or abdomen, but it never remains there, nor is the sting unsheathed, as may easily be seen. Once the struggle has commenced the Philanthus is so absorbed in her operations that I can remove the glass cover and follow every detail of the drama with my magnifying-glass.

The invariable situation of the wound being proved, I bend back the head of the bee, so as to open the articulation. I see under what we may call the chin of the bee a white spot, hardly a twenty-fifth of an inch square, where the horny integuments are lacking, and the fine skin is exposed uncovered. It is there, always there, in that tiny defect in the bee's armour, that the sting is inserted. Why is this point attacked rather than another? Is it the only point that is vulnerable? Stretch open the articulation of the corselet to the rear of the first pair of legs. There you will see an area of defenceless skin, fully as delicate as that of the throat, but much more extensive. The horny armour of the bee has no larger breach. If the Philanthus were guided solely by considerations of vulnerability she would certainly strike there, instead of insistently seeking the narrow breach in the throat. The sting would not grope or hesitate, it would find its mark at the first attempt. No; the poisoned thrust is not conditioned by mechanical considerations; the murderer disdains the wide breach in the corselet and prefers the lesser one beneath the chin, for purely logical reasons which we will now attempt to elicit.

The moment the bee is stung I release it from the aggressor. I am struck in the first place by the sudden inertia of the antennae and the various members of the mouth; organs which continue to move for so long a time in the victims of most predatory creatures. I see none of the indications with which my previous studies of paralysed victims have made me familiar: the antennae slowly waving, the mandibles opening and closing, the palpae trembling for days, for weeks, even for months. The thighs tremble for a minute or two at most; and the struggle is over. Henceforth there is complete immobility. The significance of this sudden inertia is forced upon me: the Philanthus has stabbed the cervical ganglions. Hence the sudden immobility of all the organs of the head: hence the real, not the apparent death of the bee. The Philanthus does not paralyse merely, but kills.

This is one step gained. The murderer chooses the point below the chin as the point of attack, in order to reach the principal centres of innervation, the cephalic ganglions, and thus to abolish life at a single blow. The vital centres being poisoned, immediate death must follow. If the object of the Philanthus were merely to cause paralysis she would plunge her sting into the defective corselet, as does the Cerceris in attacking the weevil, whose armour is quite unlike the bee's. Her aim is to kill outright, as we shall presently see; she wants a corpse, not a paralytic. We must admit that her technique is admirable; our human murderers could do no better.

Her posture of attack, which is very different to that of the paralysers, is infallibly fatal to the victim. Whether she delivers the attack in the erect position or prone, she holds the bee before her, head to head and thorax to thorax. In this position it suffices to flex the abdomen in order to reach the joint of the neck, and to plunge the sting obliquely upwards into the head of the captive. If the bee were seized in the inverse position, or if the sting were to go slightly astray, the results would be totally different; the sting, penetrating the bee in a downward direction, would poison the first thoracic ganglion and provoke a partial paralysis only. What art, to destroy a miserable bee! In what fencing-school did the slayer learn that terrible upward thrust beneath the chin? And as she has learned it, how is it that her victim, so learned in matters of architecture, so conversant with the politics of Socialism, has so far learned nothing in her own defence? As vigorous as the aggressor, she also carries a rapier, which is even more formidable and more painful in its results--at all events, when my finger is the victim! For centuries and centuries Philanthus has stored her cellars with the corpses of bees, yet the innocent victim submits, and the annual decimation of her race has not taught her how to deliver herself from the scourge by a well-directed thrust. I am afraid I shall never succeed in understanding how it is that the assailant has acquired her genius for sudden murder while the assailed, better armed and no less powerful, uses her dagger at random, and so far without effect. If the one has learned something from the prolonged exercise of the attack, then the other should also have learned something from the prolonged exercise of defence, for attack and defence are of equal significance in the struggle for life. Among the theorists of our day, is there any so far-sighted as to be able to solve this enigma?

I will take this opportunity of presenting a second point which embarrasses me; it is the carelessness--it is worse than that--the imbecility of the bee in the presence of the Philanthus. One would naturally suppose that the persecuted insect, gradually instructed by family misfortune, would exhibit anxiety at the approach of the ravisher, and would at least try to escape. But in my bell-glasses or wire-gauze cages I see nothing of the kind. Once the first excitement due to imprisonment has passed the bee takes next to no notice of its terrible neighbour. I have seen it side by side with Philanthus on the same flower; assassin and future victim were drinking from the same goblet. I have seen it stupidly coming to inquire what the stranger might be, as the latter crouched watching on the floor. When the murderer springs it is usually upon some bee which passes before her, and throws itself, so to speak, into her clutches; either thoughtlessly or out of curiosity. There is no frantic terror, no sign of anxiety, no tendency to escape. How is it that the experience of centuries, which is said to teach so much to the lower creatures, has not taught the bee even the beginning of apine wisdom: a deep-rooted horror of the Philanthus? Does the bee count upon its sting? But the unhappy creature is no fencer; it thrusts without method, at random. Nevertheless, let us watch it at the final and fatal moment.

When the ravisher brings her sting into play the bee also uses its sting, and with fury. I see the point thrusting now in this direction, now in that; but in empty air, or grazing and slipping over the convexity of the murderer's back, which is violently flexed. These blows have no serious results. In the position assumed by the two as they struggle the abdomen of the Philanthus is inside and that of the bee outside; thus the sting of the latter has under its point only the dorsal face of the enemy, which is convex and slippery, and almost invulnerable, so well is it armoured. There is no breach there by which the sting might possibly enter; and the operation takes place with the certainty of a skilful surgeon using the lancet, despite the indignant protests of the patient.

The fatal stroke once delivered, the murderer remains for some time on the body of the victim, clasping it face to face, for reasons that we must now consider. It may be that the position is perilous for Philanthus. The posture of attack and self-protection is abandoned, and the ventral area, more vulnerable than the back, is exposed to the sting of the bee. Now the dead bee retains for some minutes the reflex use of the sting, as I know to my cost: for removing the bee too soon from the aggressor, and handling it carelessly, I have received a most effectual sting. In her long embrace of the poisoned bee, how does Philanthus avoid this sting, which does not willingly give up its life without vengeance? Are there not sometimes unexpected accidents? Perhaps.

Here is a fact which encourages me in this belief. I had placed under the bell-glass at the same time four bees and as many Eristales, in order to judge of the entomological knowledge of Philanthus as exemplified in the distinction of species. Reciprocal quarrels broke out among the heterogeneous group. Suddenly, in the midst of the tumult, the killer is killed. Who has struck the blow? Certainly not the turbulent but pacific Eristales; it was one of the bees, which by chance had thrust truly in the mellay. When and how? I do not know. This accident is unique in my experience; but it throws a light upon the question. The bee is capable of withstanding its adversary; it can, with a thrust of its envenomed needle, kill the would-be killer. That it does not defend itself more skilfully when it falls into the hands of its enemy is due to ignorance of fencing, not to the weakness of the arm. And here again arises, more insistently than before, the question I asked but now: how is it that the Philanthus has learned for purposes of attack what the bee has not learned for purposes of defence. To this difficulty I see only one reply: the one knows without having learned and the other does not know, being incapable of learning.

Let us now examine the motives which induce the Philanthus to kill its bee instead of paralysing it. The murder once committed, it does not release its victim for a moment, but holding it tightly clasped with its six legs pressed against its body, it commences to ravage the corpse. I see it with the utmost brutality rooting with its mandibles in the articulation of the neck, and often also in the more ample articulation of the corselet, behind the first pair of legs; perfectly aware of the fine membrane in that part, although it does not take advantage of the fact when employing its sting, although this vulnerable point is the more accessible of the two breaches in the bee's armour. I see it squeezing the bee's stomach, compressing it with its own abdomen, crushing it as in a vice. The brutality of this manipulation is striking; it shows that there is no more need of care and skill. The bee is a corpse, and a little extra pushing and squeezing will not deteriorate its quality as food, provided there is no effusion of blood; and however rough the treatment, I have never been able to discover the slightest wound.

These various manipulations, above all the compression of the throat, lead to the desired result: the honey in the stomach of the bee ascends to the mouth. I see the drops of honey welling out, lapped up by the glutton as soon as they appear. The bandit greedily takes in its mouth the extended and sugared tongue of the dead insect; then once more it presses the neck and the thorax, and once more applies the pressure of its abdomen to the honey-sac of the bee. The honey oozes forth and is instantly licked up. This odious meal at the expense of the corpse is taken in a truly sybaritic attitude: the Philanthus lies upon its side with the bee between its legs. This atrocious meal lasts often half an hour and longer. Finally the exhausted corpse is abandoned; regretfully, it seems, for from time to time I have seen the ogre return to the feast and repeat its manipulation of the body. After taking a turn round the top of the bell-glass the robber of the dead returns to the victim, squeezes it once more, and licks its mouth until the last trace of honey has disappeared.

The frantic passion of the Philanthus for the honey of the bee is betrayed in another fashion. When the first victim has been exhausted I have introduced a second bee, which has been promptly stabbed under the chin and squeezed as before in order to extract its honey. A third has suffered the same fate without appeasing the bandit. I have offered a fourth, a fifth; all are accepted. My notes record that a Philanthus sacrificed six bees in succession before my eyes, and emptied them all of honey in the approved manner. The killing came to an end not because the glutton was satiated, but because my functions as provider were becoming troublesome; the dry month of August leaves but few insects in the flowerless garden. Six bees emptied of their honey--what a gluttonous meal! Yet the famishing creature would doubtless have welcomed a copious addition thereto had I had the means of furnishing it!

We need not regret the failure of bees upon this occasion; for what I have already written is sufficient testimony of the singular habits of this murderer of bees. I am far from denying that the Philanthus has honest methods of earning its living; I see it among the flowers, no less assiduous than the rest of the Hymenoptera, peacefully drinking from their cups of nectar. The male, indeed, being stingless, knows no other means of supporting himself. The mothers, without neglecting the flowers as a general thing, live by brigandage as well. It is said of the Labba, that pirate of the seas, that it pounces upon sea-birds as they rise from the waves with captured fish in their beaks. With a blow of the beak delivered in the hollow of the stomach, the aggressor forces the victim to drop its prey, and promptly catches it as it falls. The victim at least escapes with nothing worse than a blow at the base of the neck. The Philanthus, less scrupulous, falls upon the bee, stabs it to death and makes it disgorge in order to nourish herself upon its honey.

Nourish, I say, and I do not withdraw the expression. To support my statement I have better reasons than those already presented. In the cages in which various predatory Hymenoptera whose warlike habits I am studying are confined, waiting until I have procured the desired prey--not always an easy proceeding--I have planted a few heads of flowers and a couple of thistle-heads sprinkled with drops of honey, renewed at need. On these my captives feed. In the case of the Philanthus the honeyed flowers, although welcomed, are not indispensable. It is enough if from time to time I place in the cage a few living bees. Half a dozen a day is about the proper allowance. With no other diet than the honey extracted from their victims I keep my specimens of Philanthus for a fortnight and three weeks.

So much is plain: in a state of freedom, when occasion offers, the Philanthus must kill on her own account as she does in captivity. The Odynerus asks nothing of the Chrysomela but a simple condiment, the aromatic juice of the anal pouch; the Philanthus demands a full diet, or at least a notable supplement thereto, in the form of the contents of the stomach. What a hecatomb of bees must not a colony of these pirates sacrifice for their personal consumption, to say nothing of their stores of provisions! I recommend the Philanthus to the vengeance of apiarists.

For the moment we will not look further into the original causes of the crime. Let us consider matters as we know them, with all their real or apparent atrocity. In order to nourish herself the Philanthus levies tribute upon the crop of the bee. This being granted, let us consider the method of the aggressor more closely. She does not paralyse its captives according to the customary rites of the predatory insects; she kills them. Why? To the eyes of understanding the necessity of a sudden death is as clear as day. Without eviscerating the bee, which would result in the deterioration of its flesh considered as food for the larvae; without having recourse to the bloody extirpation of the stomach, the Philanthus intends to obtain its honey. By skilful manipulation, by cunning massage, she must somehow make the bee disgorge. Suppose the bee stung in the rear of the corselet and paralysed. It is deprived of locomotion, but not of vitality. The digestive apparatus, in particular, retains in full, or at least in part, its normal energies, as is proved by the frequent dejections of paralysed victims so long as the intestine is not emptied; a fact notably exemplified by the victims of the Sphex family; helpless creatures which I have before now kept alive for forty days with the aid of a little sugared water. Well! without therapeutic means, without emetics or stomach-pumps, how is a stomach intact and in good order to be persuaded to yield up its contents? That of the bee, jealous of its treasure, will lend itself to such treatment less readily than another. Paralysed, the creature is inert; but there are always internal energies and organic resistances which will not yield to the pressure of the manipulator. In vain would the Philanthus gnaw at the throat and squeeze the flanks; the honey would not return to the mouth as long as a trace of life kept the stomach closed.

Matters are different with a corpse. The springs relax; the muscles yield; the resistance of the stomach ceases, and the vessels containing the honey are emptied by the pressure of the thief. We see, therefore, that the Philanthus is obliged to inflict a sudden death which instantly destroys the contractile power of the organs. Where shall the deadly blow be delivered? The slayer knows better than we, when she pierces the victim beneath the chin. Through the narrow breach in the throat the cerebral ganglions are reached and immediate death ensues.

The examination of these acts of brigandage is not sufficient in view of my incorrigible habit of following every reply by another query, until the granite wall of the unknowable rises before me. Although the Philanthus is skilled in forcing the bee to disgorge, in emptying the crop distended with honey, this diabolical skill cannot be merely an alimentary resource, above all when in common with other insects she has access to the refectory of the flowers. I cannot regard her talents as inspired solely by the desire of a meal obtained by the labour of emptying the stomach of another insect. Something must surely escape us here: the real reason for emptying the stomach. Perhaps a respectable reason is concealed by the horrors I have recorded. What is it?

Every one will understand the vagueness which fills the observer's mind in respect of such a question as this. The reader has the right to be doubtful. I will spare him my suspicions, my gropings for the truth, and the checks encountered in the search, and give him the results of my long inquiry. Everything has its appropriate and harmonious reason. I am too fully persuaded of this to believe that the Philanthus commits her profanation of corpses merely to satisfy her appetite. What does the empty stomach mean? May it not--Yes!--But, after all, who knows? Well, let us follow up the scent.

The first care of the mothers is the welfare of the family. So far all we know of the Philanthus concerns her talent for murder. Let us consider her as a mother. We have seen her hunt on her own account; let us now watch her hunt for her offspring, for the race. Nothing is simpler than to distinguish between the two kinds of hunting. When the insect wants a few good mouthfuls of honey and nothing else, she abandons the bee contemptuously when she has emptied its stomach. It is so much valueless waste, which will shrivel where it lies and be dissected by ants. If, on the other hand, she intends to place it in the larder as a provision for her larvae, she clasps it with her two intermediate legs, and, walking on the other four, drags it to and fro along the edge of the bell-glass in search of an exit so that she may fly off with her prey. Having recognised the circular wall as impassable, she climbs its sides, now holding the bee in her mandibles by the antennae, clinging as she climbs to the vertical polished surface with all six feet. She gains the summit of the glass, stays for a little while in the flask-like cavity of the terminal button or handle, returns to the ground, and resumes her circuit of the glass and her climbing, relinquishing the bee only after an obstinate attempt to escape with it. The persistence with which the Philanthus retains her clasp upon the encumbering burden shows plainly that the game would go straight to the larder were the insect at liberty.

Those bees intended for the larvae are stung under the chin like the others; they are true corpses; they are manipulated, squeezed, exhausted of their honey, just as the others. There is no difference in the method of capture nor in their after-treatment.

As captivity might possibly result in a few anomalies of action, I decided to inquire how matters went forward in the open. In the neighbourhood of some colonies of Philanthidae I lay in wait, watching for perhaps a longer time than the question justified, as it was already settled by what occurred in captivity. My scrupulous watching at various times was rewarded. The majority of the hunters immediately entered their nests, carrying the bees pressed against their bodies; some halted on the neighbouring undergrowth; and these I saw treating the bee in the usual manner, and lapping the honey from its mouth. After these preparations the corpse was placed in the larder. All doubt was thus destroyed: the bees provided for the larvae are previously carefully emptied of their honey.

Since we are dealing with the subject, let us take the opportunity of inquiring into the customs of the Philanthus in a state of freedom. Making use of her victims when absolutely lifeless, so that they would putrefy in the course of a few days, this hunter of bees cannot adopt the customs of certain insects which paralyse their prey, and fill their cellars before laying an egg. She must surely be obliged to follow the method of the Bembex, whose larva receives, at intervals, the necessary nourishment; the amount increasing as the larva grows. The facts confirm this deduction. I spoke just now of the tediousness of my watching when watching the colonies of the Philanthus. It was perhaps even more tedious than when I was keeping an eye upon the Bembex. Before the burrows of **Cerceris tuberculus** and other devourers of the weevil, and before that of the yellow-winged Sphex, the slayer of crickets, there is plenty of distraction, owing to the busy movements of the community. The mothers have scarcely entered the nest before they are off again, returning quickly with fresh prey, only to set out once more. The going and coming is almost continuous until the storehouse is full.

The burrows of the Philanthus know nothing of such animation, even in a populous colony. In vain my vigils prolonged themselves into whole mornings or afternoons, and only very rarely does the mother who has entered with a bee set forth upon a second expedition. Two captures by the same huntress is the most that I have seen in my long watches. Once the family is provided with sufficient food for the moment the mother postpones further hunting trips until hunting becomes necessary, and busies herself with digging and burrowing in her underground dwelling. Little cells are excavated, and I see the rubbish from them gradually pushed up to the surface. With that exception there is no sign of activity; it is as though the burrow were deserted.

To lay the nest bare is not easy. The burrow penetrates to a depth of about three feet in a compact soil; sometimes in a vertical, sometimes in a horizontal direction. The spade and pick, wielded by hands more vigorous but less expert than my own, are indispensable; but the conduct of the excavation is anything but satisfactory. At the extremity of the long gallery--it seems as though the straw I use for sounding would never reach the end--we finally discover the cells, egg-shaped cavities with the longer axis horizontal. Their number and their mutual disposition escape me.

Some already contain the cocoon--slender and translucid, like that of the Cerceris, and, like it, recalling the shape of certain homoeopathic phials, with oval bodies surmounted by a tapering neck. By the extremity of the neck, which is blackened and hardened by the dejecta of the larvae, the cocoon is fixed to the end of the cell without any other support. It reminds one of a short club, planted by the end of the handle, in a line with the horizontal axis of the cell. Other cells contain the larva in a stage more or less advanced. The grub is eating the last victim proffered;

around it lie the remains of food already consumed. Others, again, show me a bee, a single bee, still intact, and having an egg deposited on the under-side of the thorax. This bee represents the first instalment of rations; others will follow as the grub matures. My expectations are thus confirmed; as with Bembex, slayer of Diptera, so Philanthus, killer of bees, lays her egg upon the first body stored, and completes, at intervals, the provisioning of the cells.

The problem of the dead bee is elucidated; there remains the other problem, of incomparable interest--Why, before they are given over to the larvae, are the bees robbed of their honey? I have said, and I repeat, that the killing and emptying of the bee cannot be explained solely by the gluttony of the Philanthus. To rob the worker of its booty is nothing; such things are seen every day; but to slaughter it in order to empty its stomach--no, gluttony cannot be the only motive. And as the bees placed in the cells are squeezed dry no less than the others, the idea occurs to me that as a beefsteak garnished with **confitures** is not to every one's taste, so the bee sweetened with honey may well be distasteful or even harmful to the larvae of the Philanthus. What would the grub do if, replete with blood and flesh, it were to find under its mandibles the honey-bag of the bee?--if, gnawing at random, it were to open the bees stomach and so drench its game with syrup? Would it approve of the mixture? Would the little ogre pass without repugnance from the gamey flavour of a corpse to the scent of flowers? To affirm or deny is useless. We must see. Let us see.

I take the young larvae of the Philanthus, already well matured, but instead of serving them with the provisions buried in their cells I offer them game of my own catching--bees that have filled themselves with nectar among the rosemary bushes. My bees, killed by crushing the head, are thankfully accepted, and at first I see nothing to justify my suspicions. Then my nurslings languish, show themselves disdainful of their food, give a negligent bite here and there, and finally, one and all, die beside their uncompleted meal. All my attempts miscarry; not once do I succeed in rearing my larvae as far as the stage of spinning the cocoon. Yet I am no novice in my duties as dry-nurse. How many pupils have passed through my hands and have reached the final stage in my old sardine-boxes as well as in their native burrows! I shall draw no conclusions from this check, which my scruples may attribute to some unknown cause. Perhaps the atmosphere of my cabinet and the dryness of the sand serving them for a bed have been too much for my nurslings, whose tender skins are used to the warm moisture of the subsoil. Let us try another method.

To decide positively whether honey is or is not repugnant to the grubs of the Philanthus was hardly practicable by the method just explained. The first meals consisted of flesh, and after that nothing in particular occurred. The honey is encountered later, when the bee is largely consumed. If hesitation and repugnance were manifested at this point they came too late to be conclusive; the sickness of the larvae might be due to other causes, known or unknown. We must offer honey at the very beginning, before artificial rearing has spoilt the grub's appetite. To offer pure honey would, of course, be useless; no carnivorous creature would touch it, even were it starving. I must spread the honey on meat; that is, I must smear the dead bee with honey, lightly varnishing it with a camel's-hair brush.

Under these conditions the problem is solved with the first few mouthfuls. The grub, having bitten on the honeyed bee, draws back as though disgusted; hesitates for a long time; then, urged by hunger, begins again; tries first on one side, then on another; in the end it refuses to touch the bee again. For a few days it pines upon its rations, which are almost intact, then dies. As many as are subjected to the same treatment perish in the same way.

Do they simply die of hunger in the presence of food which their appetites reject, or are they poisoned by the small amount of honey absorbed at the first bites? I cannot say; but, whether poisonous or merely repugnant, the bee smeared with honey is always fatal to them; a fact which explains more clearly than the unfavourable circumstances of the former experiment my lack of success with the freshly killed bees.

This refusal to touch honey, whether poisonous or repugnant, is connected with principles of alimentation too general to be a gastronomic peculiarity of the Philanthus grub. Other carnivorous larvae--at least in the series of the Hymenoptera--must share it. Let us experiment. The method need not be changed. I exhume the larvae when in a state of medium growth, to avoid the vicissitudes of extreme youth; I collect the bodies of the grubs and insects which form their natural diet and smear each body with honey, in which condition I return them to the larvae. A distinction is apparent: all the larvae are not equally suited to my experiment. Those larvae must be rejected which are nourished upon one single corpulent insect, as is that of the Scolia. The grub attacks its prey at a determined point, plunges its head and neck into the body of the insect, skilfully divides the entrails in order to keep the remains fresh until its meal is ended, and does not emerge from the opening until all is consumed but the empty skin.

To interrupt the larva with the object of smearing the interior of its prey with honey is doubly objectionable; I might extinguish the lingering vitality which keeps putrefaction at bay in the victim, and I might confuse the delicate art of the larva, which might not be able to recover the lode at which it was working or to distinguish between those parts which are lawfully and properly eaten and those which must not be consumed until a later period. As I have shown in a previous volume, the grub of the Scolia has taught me much in this respect. The only larvae acceptable for this experiment are those which are fed on a number of small insects, which are attacked without any special art, dismembered at random, and quickly consumed. Among such larvae I have experimented with those provided by chance--those of various Bembeces, fed on Diptera; those of the Palaris, whose diet consists of a large variety of Hymenoptera; those of the Tachytus, provided with young crickets; those of the Odynerus, fed upon larvae of the Chrysomela; those of the sand-dwelling Cerceris, endowed with a hecatomb of weevils. As will be seen, both consumers and consumed offer plenty of variety. Well, in every case their proper diet, seasoned with honey, is fatal. Whether poisoned or disgusted, they all die in a few days.

A strange result! Honey, the nectar of the flowers, the sole diet of the apiary under its two forms and the sole nourishment of the predatory insect in its adult phase, is for the larva of the same insect an object of insurmountable disgust, and probably a poison. The transfiguration of the chrysalis surprises me less than this inversion of the appetite. What change occurs in the stomach of the insect that the adult should passionately seek that which the larva refuses under peril of death? It is no question of organic debility unable to support a diet too substantial, too hard, or too highly spiced. The grubs which consume the larva of the Cetoniae, for example (the Rose-chafers), those which feed upon the leathery cricket, and those whose diet is rich in nitrobenzine, must assuredly have complacent gullets and adaptable stomachs. Yet these robust eaters die of hunger or poison for no greater cause than a drop of syrup, the lightest diet imaginable, adapted to the weakness of extreme youth, and a delicacy to the adult! What a gulf of obscurity in the stomach of a miserable worm!

These gastronomic experiments called for a counter-proof. The carnivorous grub is killed by honey. Is the honey-fed grub, inversely, killed by carnivorous diet? Here, again, we must make certain exceptions, observe a certain choice, as in the previous experiments. It would obviously be courting a flat refusal to offer a heap of young crickets to the larvae of the Anthophorus and the Osmia, for example; the honey-fed grub would not bite such food. It would be absolutely useless to make such an experiment. We must find the equivalent of the bee smeared with honey; that is, we must offer the larva its ordinary food with a mixture of animal matter added. I shall experiment with albumen, as provided by the egg of the hen; albumen being an isomer of fibrine, which is the principal element of all flesh diet.

Osmia tricornis will lend itself to my experiment better than any other insect on account of its dry honey, or bee-bread, which is largely formed of flowery pollen. I knead it with the albumen, graduating the dose of the latter so that its weight largely exceeds that of the bee-bread. Thus I obtain pastes of various degrees of consistency, but all firm enough to support the larva without danger of immersion. With too fluid a mixture there would be a danger of death by drowning. Finally, on each cake of albuminous paste I install a larva of medium growth.

This diet is not distasteful; far from it. The grubs attack it without hesitation and devour it with every appearance of a normal appetite. Matters could not go better if the food had not been modified according to my recipes. All is eaten; even the portions which I feared contained an excessive proportion of albumen. Moreover--a matter of still greater importance--the larvae of the Osmia fed in this manner attain their normal growth and spin their cocoons, from which adults issue in the following year. Despite the albuminous diet the cycle of evolution completes itself without mishap.

What are we to conclude from all this? I confess I am embarrassed. ***Omne vivum ex ovo***, says the physiologist. All animals are carnivorous in their first beginnings; they are formed and nourished at the expense of the egg, in which albumen predominates. The highest, the mammals, adhere to this diet for a considerable time; they live by the maternal milk, rich in casein, another isomer of albumen. The gramnivorous nestling is fed first upon worms and grubs, which are best adapted to the delicacy of its stomach; many newly born creatures among the lower orders, being immediately left to their own devices, live on animal diet. In this way the original method of alimentation is continued--the method which builds flesh out of flesh and makes blood out of blood with no chemical processes but those of simple reconstruction. In maturity, when the stomach is more robust, a vegetable diet may be adopted, involving a more complex chemistry, although the food itself is more easily obtained. To milk succeeds fodder; to the worm, seeds and grain; to the dead or paralysed insects of the natal burrow, the nectar of flowers.

Here is a partial explanation of the double system of the Hymenoptera with their carnivorous larvae--the system of dead or paralysed insects followed by honey. But here the point of interrogation, already encountered elsewhere, erects itself once again. Why is the larva of the Osmia, which thrives upon albumen, actually fed upon honey during its early life? Why is a vegetable diet the rule in the hives of bees from the very commencement, when the other members of the same series live upon animal food?

If I were a "transformist" how I should delight in this question! Yes, I should say: yes, by the fact of its germ every animal is originally carnivorous. The insect in particular makes a beginning with albuminoid materials. Many larvae adhere to the alimentation present in the egg, as do many adult insects also. But the struggle to fill the belly, which is actually the struggle for life, demands something better than the precarious chances of the chase. Man, at first an eager hunter of game, collected flocks and became a shepherd in order to profit by his possessions in time of dearth. Further progress inspired him to till the earth and sow; a method which assured him of a certain living. Evolution from the defective to the mediocre, and from the mediocre to the abundant, has led to the resources of agriculture.

The lower animals have preceded us on the way of progress. The ancestors of the Philanthus, in the remote ages of the lacustrian tertiary formations, lived by capturing prey in both phases--both as larvae and as adults; they hunted for their own benefit as well as for the family. They did not confine themselves to emptying the stomach of the bee, as do their descendants to-day; they devoured the victim entire. From beginning to end they remained carnivorous. Later there were fortunate innovators, whose race supplanted the more conservative element, who discovered an inexhaustible source of nourishment, to be obtained without painful search or dangerous conflict: the saccharine exudation of the flowers. The wasteful system of living upon prey, by no means favourable to large populations, has been preserved for the feeble larvae; but the vigorous adult has abandoned it for an easier and more prosperous existence. Thus the Philanthus of our own days was gradually developed; thus was formed the double system of nourishment practised by the various predatory insects which we know.

The bee has done still better; from the moment of leaving the egg it dispenses completely with chance-won aliments. It has invented honey, the food of its larvae. Renouncing the chase for ever, and becoming exclusively agricultural, this insect has acquired a degree of moral and physical prosperity that the predatory species are far from sharing. Hence the flourishing colonies of the Anthophorae, the Osmiae, the Eucerae, the Halicti, and other makers of honey, while the hunters of prey work in isolation; hence the societies in which the bee displays its admirable talents, the supreme expression of instinct.

This is what I should say if I were a "transformist." All this is a chain of highly logical deductions, and it hangs together with a certain air of reality, such as we like to look for in a host of "transformist" arguments which are put forward as irrefutable. Well, I make a present of my deductive theory to whosoever desires it, and without the least regret; I do not believe a single word of it, and I confess my profound ignorance of the origin of the twofold system of diet.

One thing I do see more clearly after all my experiments and research: the tactics of the Philanthus. As a witness of its ferocious feasting, the true motive of which was unknown to me, I treated it to all the unfavourable epithets I could think of; called it assassin, bandit, pirate, robber of the dead. Ignorance is always abusive; the man who does not know is full of violent affirmations and malign interpretations. Undeceived by the facts, I hasten to apologise and express my esteem for the Philanthus. In emptying the stomach of the bee the mother is performing the most praiseworthy of all duties; she is guarding her family against poison. If she sometimes kills on her own account and abandons the body after exhausting it of honey, I dare not call her action a crime. When the habit has once been formed of emptying the bee's crop for the best of motives, the temptation is great to do so with no other excuse than hunger. Moreover--who can say?--perhaps there is always some afterthought that the larvae might profit by the sacrifice. Although not carried into effect the intention excuses the act.

I therefore withdraw my abusive epithets in order to express my admiration of the creature's maternal logic. Honey would be harmful to the grubs. How does the mother know that honey, in which she herself delights, is noxious to her young? To this question our knowledge has no reply. But honey, as we have seen, would endanger the lives of the grubs. The bees must therefore be emptied of honey before they are fed to them. The process must be effected without wounding the victim, for the larva must receive the latter fresh and moist; and this would be impracticable if the insect were paralysed on account of the natural resistance of the organs. The bee must therefore be killed outright instead of being paralysed, otherwise the honey could not be removed. Instantaneous death can be assured only by a lesion of the primordial centre of life. The sting must therefore pierce the cervical ganglions; the centre of innervation upon which the rest of the organism is dependent. This can only be reached in one way: through the neck. Here it is that the sting will be inserted; and here it is inserted in a breach in the armour no larger than a pin's head. Suppress a single link of this closely knit chain, and the Philanthus reared upon the flesh of bees becomes an impossibility.

That honey is fatal to larvae is a fact pregnant with consequences. Various predatory insects feed their young with honey-makers. Such, to my knowledge, are the *Philanthus coronatus*, Fabr., which stores its burrows with the large Halictus; the *Philanthus raptor*, Lep., which chases all the smaller Halictus indifferently, being itself a small insect; the *Cerceris ornata*, Fabr., which also kills Halictus; and the *Polaris flavipes*, Fabr., which by a strange eclecticism fills its cells with specimens of most of the Hymenoptera which are not beyond its powers. What do these four huntresses, and others of similar habits, do with their victims when the crops of the latter are full of honey? They must follow the example of the Philanthus or their offspring would perish; they must squeeze and manipulate the dead bee until it yields up its honey. Everything goes to prove as much; but for the actual observation of what would be a notable proof of my theory I must trust to the future.

CHAPTER XIV
THE GREAT PEACOCK, OR EMPEROR MOTH

It was a memorable night! I will name it the Night of the Great Peacock. Who does not know this superb moth, the largest of all our European butterflies[3] with its livery of chestnut velvet and its collar of white fur? The greys and browns of the wings are crossed by a paler zig-zag, and bordered with smoky white; and in the centre of each wing is a round spot, a great eye with a black pupil and variegated iris, resolving into concentric arcs of black, white, chestnut, and purplish red.

Not less remarkable is the caterpillar. Its colour is a vague yellow. On the summit of thinly sown tubercles crowned with a palisade of black hairs are set pearls of a turquoise-blue. The burly brown cocoon, which is notable for its curious tunnel of exit, like an eel-pot, is always found at the base of an old almond-tree, adhering to the bark. The foliage of the same tree nourishes the caterpillar.

On the morning of the 6th of May a female emerged from her cocoon in my presence on my laboratory table. I cloistered her immediately, all damp with the moisture of metamorphosis, in a cover of wire gauze. I had no particular intentions regarding her; I imprisoned her from mere habit; the habit of an observer always on the alert for what may happen.

I was richly rewarded. About nine o'clock that evening, when the household was going to bed, there was a sudden hubbub in the room next to mine. Little Paul, half undressed, was rushing to and fro, running, jumping, stamping, and overturning the chairs as if possessed. I heard him call me. "Come quick!" he shrieked; "come and see these butterflies! Big as birds! The room's full of them!"

I ran. There was that which justified the child's enthusiasm and his hardly hyperbolical exclamation. It was an invasion of giant butterflies; an invasion hitherto unexampled in our house. Four were already caught and placed in a bird-cage. Others--numbers of them--were flying across the ceiling.

This astonishing sight recalled the prisoner of the morning to my mind. "Put on your togs, kiddy!" I told my son; "put down your cage, and come with me. We shall see something worth seeing."

We had to go downstairs to reach my study, which occupies the right wing of the house. In the kitchen we met the servant; she too was bewildered by the state of affairs. She was pursuing the huge butterflies with her apron, having taken them at first for bats.

It seemed as though the Great Peacock had taken possession of my whole house, more or less. What would it be upstairs, where the prisoner was, the cause of this invasion? Happily one of the two study windows had been left ajar; the road was open.

Candle in hand, we entered the room. What we saw is unforgettable. With a soft *flic-flac* the great night-moths were flying round the wire-gauze cover, alighting, taking flight, returning, mounting to the ceiling, re-descending. They rushed at the candle and extinguished it with a flap of the wing; they fluttered on our shoulders, clung to our clothing, grazed our faces. My study had become a cave of a necromancer, the darkness alive with creatures of the night! Little Paul, to reassure himself, held my hand much tighter than usual.

How many were there? About twenty. To these add those which had strayed into the kitchen, the nursery, and other rooms in the house, and the total must have been nearly forty. It was a memorable sight--the Night of the Great Peacock! Come from all points of the compass, warned I know not how, here were forty lovers eager to do homage to the maiden princess that morning born in the sacred precincts of my study.

For the time being I troubled the swarm of pretenders no further. The flame of the candle endangered the visitors; they threw themselves into it stupidly and singed themselves slightly. On the morrow we could resume our study of them, and make certain carefully devised experiments.

To clear the ground a little for what is to follow, let me speak of what was repeated every night during the eight nights my observations lasted. Every night, when it was quite dark, between eight and ten o'clock, the butterflies arrived one by one. The weather was stormy; the sky heavily clouded; the darkness was so profound that out of doors, in the garden and away from the trees, one could scarcely see one's hand before one's face.

In addition to such darkness as this there were certain difficulties of access. The house is hidden by great plane-trees; an alley densely bordered with lilacs and rose-trees make a kind of outer vestibule to the entrance; it is protected from the *mistral* by groups of pines and screens of cypress. A thicket of evergreen shrubs forms a rampart at a few paces from the door. It was across this maze of leafage, and in absolute darkness, that the butterflies had to find their way in order to attain the end of their pilgrimage.

Under such conditions the screech-owl would not dare to forsake its hollow in the olive-tree. The butterfly, better endowed with its faceted eyes than the owl with its single pupils, goes forward without hesitation, and threads the obstacles without contact. So well it directs its tortuous flight that, in spite of all the obstacles to be evaded, it arrives in a state of perfect freshness, its great wings intact, without the slightest flaw. The darkness is light enough for the butterfly.

Even if we suppose it to be sensitive to rays unknown to the ordinary retina, this extraordinary sight could not be the sense that warns the butterfly at a distance and brings it hastening to the bride. Distance and the objects interposed make the suggestion absurd.

Moreover, apart from illusory refractions, of which there is no question here, the indications of light are precise; one goes straight to the object seen. But the butterfly was sometimes mistaken: not in the general direction, but concerning the precise position of the attractive object. I have mentioned that the nursery on the other side of the house to my study, which was the actual goal of the visitors, was full of butterflies before a light was taken into it. These were certainly incorrectly informed. In the kitchen there was the same crowd of seekers gone astray; but there the light of a lamp, an irresistible attraction to nocturnal insects, might have diverted the pilgrims.

Let us consider only such areas as were in darkness. There the pilgrims were numerous. I found them almost everywhere in the neighbourhood of their goal. When the captive was in my study the butterflies did not all enter by the open window, the direct and easy way, the captive being only a few yards from the window.

Several penetrated the house downstairs, wandered through the hall, and reached the staircase, which was barred at the top by a closed door.

These data show us that the visitors to the wedding-feast did not go straight to their goal as they would have done were they attracted by any kind of luminous radiations, whether known or unknown to our physical science. Something other than radiant energy warned them at a distance, led them to the neighbourhood of the precise spot, and left the final discovery to be made after a vague and hesitating search. The senses of hearing and smell warn us very much in this way; they are not precise guides when we try to determine exactly the point of origin of a sound or smell.

What sense is it that informs this great butterfly of the whereabouts of his mate, and leads him wandering through the night? What organ does this sense affect? One suspects the antennae; in the male butterfly they actually seem to be sounding, interrogating empty space with their long feathery plumes. Are these splendid plumes merely items of finery, or do they really play a part in the perception of the effluvia which guide the lover? It seemed easy, on the occasion I spoke of, to devise a conclusive experiment.

On the morrow of the invasion I found in my study eight of my nocturnal visitors. They were perched, motionless, upon the cross-mouldings of the second window, which had remained closed. The others, having concluded their ballet by about ten o'clock at night, had left as they had entered, by the other window, which was left open night and day. These eight persevering lovers were just what I required for my experiment.

With a sharp pair of scissors, and without otherwise touching the butterflies, I cut off their antennae near the base. The victims barely noticed the operation. None moved; there was scarcely a flutter of the wings. Their condition was excellent; the wound did not seem to be in the least serious. They were not perturbed by physical suffering, and would therefore be all the better adapted to my designs. They passed the rest of the day in placid immobility on the cross-bars of the window.

A few other arrangements were still to be made. In particular it was necessary to change the scene; not to leave the female under the eyes of the mutilated butterflies at the moment of resuming their nocturnal flight; the difficulty of the search must not be lessened. I therefore removed the cage and its captive, and placed it under a porch on the other side of the house, at a distance of some fifty paces from my study.

At nightfall I went for a last time to inspect my eight victims. Six had left by the open window; two still remained, but they had fallen on the floor, and no longer had the strength to recover themselves if turned over on their backs. They were exhausted, dying. Do not accuse my surgery, however. Such early decease was observed repeatedly, with no intervention on my part.

Six, in better condition, had departed. Would they return to the call that attracted them the night before? Deprived of their antennae, would they be able to find the captive, now placed at a considerable distance from her original position?

The cage was in darkness, almost in the open air. From time to time I visited it with a net and lantern. The visitors were captured, inspected, and immediately released in a neighbouring room, of which I closed the door. This gradual elimination allowed me to count the visitors exactly without danger of counting the same butterfly more than once. Moreover, the provisional prison, large and bare, in no wise harmed or endangered the prisoners; they found a quiet retreat there and ample space. Similar precautions were taken during the rest of my experiments.

After half-past ten no more arrived. The reception was over. Total, twenty-five males captured, of which one only was deprived of its antennae. So of the six operated on earlier in the day, which were strong enough to leave my study and fly back to the fields, only one had returned to the cage. A poor result, in which I could place no confidence as proving whether the antennae did or did not play a directing part. It was necessary to begin again upon a larger scale.

Next morning I visited the prisoners of the day before. What I saw was not encouraging. A large number were scattered on the ground, almost inert. Taken between the fingers, several of them gave scarcely a sign of life. Little was to be hoped from these, it would seem. Still, I determined to try; perhaps they would regain their vigour at the lover's hour.

The twenty-four prisoners were all subjected to the amputation of their antennae. The one operated on the day before was put aside as dying or nearly so. Finally the door of the prison was left open for the rest of the day. Those might leave who could; those could join in the carnival who were able. In order to put those that might leave the room to the test of a search, the cage, which they must otherwise have encountered at the threshold, was again removed, and placed in a room of the opposite wing, on the ground floor. There was of course free access to this room.

Of the twenty-four lacking their antennae sixteen only left the room. Eight were powerless to do so; they were dying. Of the sixteen, how many returned to the cage that night? Not one. My captives that night were only seven, all new-comers, all wearing antennae. This result seemed to prove that the amputation of the antennae was a matter of serious significance. But it would not do to conclude as yet: one doubt remained.

"A fine state I am in! How shall I dare to appear before the other dogs?" said Mouflard, the puppy whose ears had been pitilessly docked. Had my butterflies apprehensions similar to Master Mouflard's? Deprived of their beautiful plumes, were they ashamed to appear in the midst of their rivals, and to prefer their suits? Was it confusion on their part, or want of guidance? Was it not rather exhaustion after an attempt exceeding the duration of an ephemeral passion? Experience would show me.

On the fourth night I took fourteen new-comers and set them apart as they came in a room in which they spent the night. On the morrow, profiting by their diurnal immobility, I removed a little of the hair from the centre of the corselet or neck. This slight tonsure did not inconvenience the insects, so easily was the silky fur removed, nor did it deprive them of any organ which might later on be necessary in the search for the female. To them it was nothing; for me it was the unmistakable sign of a repeated visit.

This time there were none incapable of flight. At night the fourteen shavelings escaped into the open air. The cage, of course, was again in a new place. In two hours I captured twenty butterflies, of whom two were tonsured; no more. As for those whose antennae I had amputated the night before, not one reappeared. Their nuptial period was over.

Of fourteen marked by the tonsure two only returned. Why did the other twelve fail to appear, although furnished with their supposed guides, their antennae? To this I can see only one reply: that the Great Peacock is promptly exhausted by the ardours of the mating season.

With a view to mating, the sole end of its life, the great moth is endowed with a marvellous prerogative. It has the power to discover the object of its desire in spite of distance, in spite of obstacles. A few hours, for two or three nights, are given to its search, its nuptial flights. If it cannot profit by them, all is ended; the compass fails, the lamp expires. What profit could life hold henceforth? Stoically the creature withdraws into a corner and sleeps the last sleep, the end of illusions and the end of suffering.

The Great Peacock exists as a butterfly only to perpetuate itself. It knows nothing of food. While so many others, joyful banqueters, fly from flower to flower, unrolling their spiral trunks to plunge them into honeyed blossoms, this incomparable ascetic, completely freed from the servitude of the stomach, has no means of restoring its strength. Its buccal members are mere vestiges, useless simulacra, not real organs able to perform their duties. Not a sip of honey can ever enter its stomach; a magnificent prerogative, if it is not long enjoyed. If the lamp is to burn it must be filled with oil. The Great Peacock renounces the joys of the palate; but with them it surrenders long life. Two or three nights--just long enough to allow the couple to meet and mate--and all is over; the great butterfly is dead.

What, then, is meant by the non-appearance of those whose antennae I removed? Did they prove that the lack of antennae rendered them incapable of finding the cage in which the prisoner waited? By no means. Like those marked with the tonsure, which had undergone no damaging operation, they proved only that their time was finished. Mutilated or intact, they could do no more on account of age, and their absence meant nothing. Owing to the delay inseparable from the experiment, the part played by the antennae escaped me. It was doubtful before; it remained doubtful.

My prisoner under the wire-gauze cover lived for eight days. Every night she attracted a swarm of visitors, now to one part of the house, now to another. I caught them with the net and released them as soon as captured in a closed room, where they passed the night. On the next day they were marked, by means of a slight tonsure on the thorax.

The total number of butterflies attracted on these eight nights amounted to a hundred and fifty; a stupendous number when I consider what searches I had to undertake during the two following years in order to collect the specimens necessary to the continuation of my investigation. Without being absolutely undiscoverable, in my immediate neighbourhood the cocoons of the Great Peacock are at least extremely rare, as the trees on which they are found are not common. For two winters I visited all the decrepit almond-trees at hand, inspected them all at the base of the trunk, under the jungle of stubborn grasses and undergrowth that surrounded them; and how often I returned with empty hands! Thus my hundred and fifty butterflies had come from some little distance; perhaps from a radius of a mile and a quarter or more. How did they learn of what was happening in my study?

Three agents of information affect the senses at a distance: sight, sound, and smell. Can we speak of vision in this connection? Sight could very well guide the arrivals once they had entered the open window; but how could it help them out of doors, among unfamiliar surroundings? Even the fabulous eye of the lynx, which could see through walls, would not be sufficient; we should have to imagine a keenness of vision capable of annihilating leagues of space. It is needless to discuss the matter further; sight cannot be the guiding sense.

Sound is equally out of the question. The big-bodied creature capable of calling her mates from such a distance is absolutely mute, even to the most sensitive ear. Does she perhaps emit vibrations of such delicacy or rapidity that only the most sensitive microphone could appreciate them? The idea is barely possible; but let us remember that the visitors must have been warned at distances of some thousands of yards. Under these conditions it is useless to think of acoustics.

Smell remains. Scent, better than any other impression in the domain of our senses, would explain the invasion of butterflies, and their difficulty at the very last in immediately finding the object of their search. Are there effluvia analogous to what we call odour: effluvia of extreme subtlety, absolutely imperceptible to us, yet capable of stimulating a sense-organ far more sensitive than our own? A simple experiment suggested itself. I would mask these effluvia, stifle them under a powerful, tenacious odour, which would take complete possession of the sense-organ and neutralise the less powerful impression.

I began by sprinkling naphthaline in the room intended for the reception of the males that evening. Beside the female, inside the wire-gauze cover, I placed a large capsule full of the same substance. When the hour of the nocturnal visit arrived I had only to stand at the door of the room to smell a smell as of a gas-works. Well, my artifice failed. The butterflies arrived as usual, entered the room, traversed its gas-laden atmosphere, and made for the wire-gauze cover with the same certainty as in a room full of fresh air.

My confidence in the olfactory theory was shaken. Moreover, I could not continue my experiments. On the ninth day, exhausted by her fruitless period of waiting, the female died, having first deposited her barren eggs upon the woven wire of her cage. Lacking a female, nothing could be done until the following year.

I determined next time to take suitable precautions and to make all preparations for repeating at will the experiments already made and others which I had in mind. I set to work at once, without delay.

In the summer I began to buy caterpillars at a halfpenny apiece.

The market was in the hands of some neighbouring urchins, my habitual providers. On Friday, free of the terrors of grammar, they scoured the fields, finding from time to time the Great Peacock caterpillar, and bringing it to me clinging to the end of a stick. They did not dare to touch it, poor little imps! They were thunderstruck at my audacity when I seized it in my fingers as they would the familiar silkworm.

Reared upon twigs of the almond-tree, my menagerie soon provided me with magnificent cocoons. In winter assiduous search at the base of the native trees completed my collection. Friends interested in my researches came to my aid. Finally, after some trouble, what with an open market, commercial negotiations, and searching, at the cost of many scratches, in the undergrowth, I became the owner of an assortment of cocoons of which twelve, larger and heavier than the rest, announced that they were those of females.

Disappointment awaited me. May arrived; a capricious month which set my preparations at naught, troublesome as these had been. Winter returned. The *mistral* shrieked, tore the budding leaves of the plane-trees, and scattered them over the ground. It was cold as December. We had to light fires in the evening, and resume the heavy clothes we had begun to leave off.

My butterflies were too sorely tried. They emerged late and were torpid. Around my cages, in which the females waited--to-day one, to-morrow another, according to the order of their birth--few males or none came from without. Yet there were some in the neighbourhood, for those with large antennae which issued from my collection of cocoons were placed in the garden directly they had emerged, and were recognised. Whether neighbours or strangers, very few came, and those without enthusiasm. For a moment they entered, then disappeared and did not reappear. The lovers were as cold as the season.

Perhaps, too, the low temperature was unfavourable to the informing effluvia, which might well be increased by heat and lessened by cold as is the case with many odours. My year was lost. Research is disappointing work when the experimenter is the slave of the return and the caprices of a brief season of the year.

For the third time I began again. I reared caterpillars; I scoured the country in search of cocoons. When May returned I was tolerably provided. The season was fine, responding to my hopes. I foresaw the affluence of butterflies which had so impressed me at the outset, when the famous invasion occurred which was the origin of my experiments.

Every night, by squadrons of twelve, twenty, or more, the visitors appeared. The female, a strapping, big-bellied matron, clung to the woven wire of the cover. There was no movement on her part; not even a flutter of the wings. One would have thought her indifferent to all that occurred. No odour was emitted that was perceptible to the most sensitive nostrils of the household; no sound that the keenest ears of the household could perceive. Motionless, recollected, she waited.

The males, by twos, by threes and more, fluttered upon the dome of the cover, scouring over it quickly in all directions, beating it continually with the ends of their wings. There were no conflicts between rivals. Each did his best to penetrate the enclosure, without betraying any sign of jealousy of the others. Tiring of their fruitless attempts, they would fly away and join the dance of the gyrating crowd. Some, in despair, would escape by the open window: new-comers would replace them: and until ten o'clock or thereabouts the wire dome of the cover would be the scene of continual attempts at approach, incessantly commencing, quickly wearying, quickly resumed.

Every night the position of the cage was changed. I placed it north of the house and south; on the ground-floor and the first floor; in the right wing of the house, or fifty yards away in the left wing; in the open air, or hidden in some distant room. All these sudden removals, devised to put the seekers off the scent, troubled them not at all. My time and my pains were wasted, so far as deceiving them was concerned.

The memory of places has no part in the finding of the female. For instance, the day before the cage was installed in a certain room. The males visited the room and fluttered about the cage for a couple of hours, and some even passed the night there. On the following day, at sunset, when I moved the cage, all were out of doors. Although their lives are so ephemeral, the youngest were ready to resume their nocturnal expeditions a second and even a third time. Where did they first go, these veterans of a day?

They knew precisely where the cage had been the night before. One would have expected them to return to it, guided by memory; and that not finding it they would go out to continue their search elsewhere. No; contrary to my expectation, nothing of the kind appeared. None came to the spot which had been so crowded the night before; none paid even a passing visit. The room was recognised as an empty room, with no previous examination, such as would apparently be necessary to contradict the memory of the place. A more positive guide than memory called them elsewhere.

Hitherto the female was always visible, behind the meshes of the wire-gauze cover. The visitors, seeing plainly in the dark night, must have been able to see her by the vague luminosity of what for us is the dark. What would happen if I imprisoned her in an opaque receptacle? Would not such a receptacle arrest or set free the informing effluvia according to its nature?

Practical physics has given us wireless telegraphy by means of the Hertzian vibrations of the ether. Had the Great Peacock butterfly outstripped and anticipated mankind in this direction? In order to disturb the whole surrounding neighbourhood, to warn pretenders at a distance of a mile or more, does the newly emerged female make use of electric or magnetic waves, known or unknown, that a screen of one material would arrest while another would allow them to pass? In a word, does she, after her fashion, employ a system of wireless telegraphy? I see nothing impossible in this; insects are responsible for many inventions equally marvellous.

Accordingly I lodged the female in boxes of various materials; boxes of tin-plate, wood, and cardboard. All were hermetically closed, even sealed with a greasy paste. I also used a glass bell resting upon a base-plate of glass.

Under these conditions not a male arrived; not one, though the warmth and quiet of the evening were propitious. Whatever its nature, whether of glass, metal, card, or wood, the closed receptacle was evidently an insuperable obstacle to the warning effluvia.

A layer of cotton-wool two fingers in thickness had the same result. I placed the female in a large glass jar, and laced a piece of thin cotton batting over the mouth for a cover; this again guarded the secret of my laboratory. Not a male appeared.

But when I placed the females in boxes which were imperfectly closed, or which had chinks in their sides, or even hid them in a drawer or a cupboard, I found the males arrived in numbers as great as when the object of their search lay in the cage of open wire-work freely exposed on a table. I have a vivid memory of one evening when the recluse was hidden in a hat-box at the bottom of a wall-cupboard. The arrivals went straight to the closed doors, and beat them with their wings, *toc-toc*, trying to enter. Wandering pilgrims, come from I know not where, across fields and meadows, they knew perfectly what was behind the doors of the cupboard.

So we must abandon the idea that the butterfly has any means of communication comparable to our wireless telegraphy, as any kind of screen, whether a good or a bad conductor, completely stops the signals of the female. To give them free passage and allow them to penetrate to a distance one condition is indispensable: the enclosure in which the captive is confined must not be hermetically sealed; there must be a communication between it and the outer air. This again points to the probability of an odour, although this is contradicted by my experiment with the naphthaline.

My cocoons were all hatched, and the problem was still obscure. Should I begin all over again in the fourth year? I did not do so, for the reason that it is difficult to observe a nocturnal butterfly if one wishes to follow it in all its intimate actions. The lover needs no light to attain his ends; but my imperfect human vision cannot penetrate the darkness. I should require a candle at least, and a candle would be constantly extinguished by the revolving swarm. A lantern would obviate these eclipses, but its doubtful light, interspersed with heavy shadows, by no means commends it to the scruples of an observer, who must see, and see well.

Moreover, the light of a lamp diverts the butterflies from their object, distracts them from their affairs, and seriously compromises the success of the observer. The moment they enter, they rush frantically at the flame, singe their down, and thereupon, terrified by the heat, are of no profit to the observer. If, instead of being roasted, they are held at a distance by an envelope of glass, they press as closely as they can to the flame, and remain motionless, hypnotised.

One night, the female being in the dining-room, on the table, facing the open window, a petroleum lamp, furnished with a large reflector in opaline glass, was hanging from the ceiling. The arrivals alighted on the dome of the wire-gauze cover, crowding eagerly about the prisoner; others, saluting her in passing, flew to the lamp, circled round it a few times, and then, fascinated by the luminous splendour radiating from the opal cone of light, clung there motionless under the reflector. Already the children were raising their hands to seize them. "Leave them," I said, "leave them. Let us be hospitable: do not disturb the pilgrims who have come to the tabernacle of the light."

During the whole evening not one of them moved. Next day they were still there. The intoxication of the light had made them forget the intoxication of love.

With creatures so madly in love with the light precise and prolonged experimentation is impracticable the moment the observer requires artificial light. I renounced the Great Peacock and its nocturnal habits. I required a butterfly with different habits; equally notable as a lover, but seeking out the beloved by day.

Before going on to speak of my experiments with a subject fulfilling these conditions, let me break the chronological order of my record in order to say a few words concerning another insect, which appeared after I had completed these inquiries. I refer to the Lesser Peacock (Attacus pavonia minor, Lin.).

Some one brought me, from what locality I do not know, a superb cocoon enveloped in an ample wrapping of white silk. From this covering, which lay in large irregular folds, the chrysalis was easily detached; in shape like that of the Great Peacock, but considerably less in size. The anterior extremity, which is defended by an arrangement of fine twigs, converging, and free at the converging ends, forming a device not unlike an eel-pot, which presents access to the chrysalis while allowing the butterfly to emerge without breaking the defence, indicated a relative of the great nocturnal butterfly; the silk-work denoted a spinning caterpillar.

Towards the end of March this curious cocoon yielded up a female of the Lesser Peacock, which was immediately sequestered under a wire-gauze cover in my study. I opened the window to allow news of the event to reach the surrounding country, and left it open so that such visitors as presented themselves should find free access to the cage. The captive clung to the wire gauze and did not move for a week.

She was a superb creature, this prisoner of mine, with her suit of brown velvet, crossed by undulating lines. The neck was surrounded by white fur; there was a carmine spot at the extremity of the upper wings, and four great eyes in which were grouped, in concentric crescents, black, white, red, and yellow ochre: almost the colouring of the Great Peacock, but more vivid. Three or four times in my life I had encountered this butterfly, so remarkable for its size and its costume. The cocoon I had recently seen for the first time; the male I had never seen. I only knew that, according to the books, it was half the size of the female, and less vividly coloured, with orange-yellow on the lower wings.

Would he appear, the elegant unknown, with waving plumes; the butterfly I had never yet seen, so rare does the Lesser Peacock seem to be in our country? Would he, in some distant hedge, receive warning of the bride who waited on my study table? I dared to hope it, and I was right. He arrived even sooner than I had hoped.

Noon struck as we were sitting down to table, when little Paul, delayed by his absorption in the expected event, suddenly ran to rejoin us, his cheeks glowing. Between his fingers we saw the fluttering wings of a handsome butterfly, caught but a moment before, while it was hovering in front of my study. He showed it me, questioning me with his eyes.

"Aha!" I cried, "this is precisely the pilgrim we are waiting for. Fold your napkin and come and see what happens. We will dine later."

Dinner was forgotten before the marvels that came to pass. With inconceivable punctuality the butterflies hastened to meet the magical call of the captive. With tortuous flight they arrived one by one. All came from the north. This detail is significant. A week earlier there had been a savage return of the winter. The *bise* blew tempestuously, killing the early almond blossom. It was one of those ferocious storms which in the South commonly serve as a prelude to the spring. But the temperature had now suddenly softened, although the wind still blew from the north.

Now on this first occasion all the butterflies hastening to the prisoner entered the garden from the north. They followed the direction of the wind; not one flew against it. If their guide was a sense of smell like ours, if they were guided by fragrant atoms suspended in the air, they should have arrived in the opposite direction. Coming from the south, we might believe them to be warned by effluvia carried on the wind; coming from the north in time of **mistral**, that resistless sweeper of earth and air, how can we suppose that they had perceived, at a remote distance, what we will call an odour? The idea of a flow of odoriferous atoms in a direction contrary to that of the aerial torrent seems to me inadmissible.

For two hours, under a radiant sun, the visitors came and went before the outer wall of the study. Most of them sought for a long time, exploring the wall, flying on a level with the ground. To see them thus hesitating you would say that they were puzzled to find the exact position of the lure which called them. Although they had come from such a distance without a mistake, they seemed imperfectly informed once they were on the spot. Nevertheless, sooner or later they entered the room and saluted the captive, without showing any great ardour. At two o'clock all was over. Ten butterflies had arrived.

During the whole week, and always about noon, at the hour of the brightest sunlight, the butterflies arrived, but in decreasing numbers. The total approached forty. I thought it useless to repeat experiments which would add nothing to what I had already learned. I will confine myself to stating two facts. In the first place, the Lesser Peacock is diurnal; that is to say, it celebrates its mating under the dazzling brilliance of noon. It needs the full force of the sunlight. The Great Peacock, on the contrary, which it so closely resembles both in its adult form and the work of its caterpillar, requires the darkness of the first hours of the night. Who can explain this strange contrast in habits?

In the second place, a powerful current of air, sweeping away in a contrary direction all particles that might inform the sense of smell, does not prevent the butterflies from arriving from a direction opposite to that taken by the effluvial stream, as we understand such matters.

To continue: I needed a diurnal moth or butterfly: not the Lesser Peacock, which came too late, when I had nothing to ask of it, but another, no matter what, provided it was a prompt guest at the wedding feast. Was I to find such an insect?

CHAPTER XV
THE OAK EGGAR, OR BANDED MONK

Yes: I was to find it. I even had it already in my possession. An urchin of seven years, with an alert countenance, not washed every day, bare feet, and dilapidated breeches supported by a piece of string, who frequented the house as a dealer in turnips and tomatoes, arrived one day with his basket of vegetables. Having received the few halfpence expected by his mother as the price of the garden-stuff, and having counted them one by one into the hollow of his hand, he took from his pocket an object which he had discovered the day before beneath a hedge when gathering greenstuff for his rabbits.

"And this--will you have this?" he said, handing me the object. "Why, certainly I will have it. Try to find me more, as many as you can, and on Sunday you shall have lots of rides on the wooden horses. In the meantime here is a penny for you. Don't forget it when you make up your accounts; don't mix it with your turnip-money; put it by itself." Beaming with satisfaction at such wealth, little touzle-head promised to search industriously, already foreseeing a fortune.

When he had gone I examined the thing. It was worth examination. It was a fine cocoon, thick and with blunt ends, very like a silkworm's cocoon, firm to the touch and of a tawny colour. A brief reference to the text-books almost convinced me that this was a cocoon of the ***Bombyx quercus***.[4] If so, what a find! I could continue my inquiry and perhaps confirm what my study of the Great Peacock had made me suspect.

The Bombyx of the oak-tree is, in fact, a classic moth; indeed, there is no ento-mological text-book but speaks of its exploits at mating-time. It is said that a female emerged from the pupa in captivity, in the interior of an apartment, and even in a closed box. It was far from the country, amidst the tumult of a large city. Nevertheless, the event was known to those concerned in the woods and meadows. Guided by some mysterious compass, the males arrived, hastening from the distant fields; they went to the box, fluttered against it, and flew to and fro in the room.

These marvels I had learned by reading; but to see such a thing with one's own eyes, and at the same time to devise experiments, is quite another thing. What had my penny bargain in store for me? Would the famous Bombyx issue from it?

Let us call it by its other name, the Banded Monk. This original name of Monk was suggested by the costume of the male; a monk's robe of a modest rusty red. But in the case of the female the brown fustian gives place to a beautiful velvet, with a pale transversal band and little white eyes on the fore pair of wings.

The Monk is not a common butterfly which can be caught by any one who takes out a net at the proper season. I have never seen it around our village or in the solitude of my grounds during a residence of twenty years. It is true that I am not a fervent butterfly-catcher; the dead insect of the collector's cabinet has little interest for me; I must have it living, in the exercise of its functions. But although I have not the collector's zeal I have an attentive eye to all that flies or crawls in the fields. A butterfly so remarkable for its size and colouring would never have escaped my notice had I encountered it.

The little searcher whom I had enticed by a promise of rides upon wooden horses never made a second find. For three years I requisitioned friends and neighbours, and especially their children, sharp-sighted snappers-up of trifles; I myself hunted often under heaps of withered leaves; I inspected stone-heaps and visited hollow tree-trunks. Useless pains; the precious cocoon was not to be found. It is enough to say that the Banded Monk is extremely rare in my neighbourhood. The importance of this fact will presently appear.

As I suspected, my cocoon was truly that of the celebrated Oak Eggar. On the 20th of August a female emerged from it: corpulent, big-bellied, coloured like the male, but lighter in hue. I placed her under the usual wire cover in the centre of my laboratory table, littered as it was with books, bottles, trays, boxes, test-tubes, and other apparatus. I have explained the situation in speaking of the Great Peacock. Two windows light the room, both opening on the garden. One was closed, the other open day and night. The butterfly was placed in the shade, between the lines of the two windows, at a distance of 12 or 15 feet.

The rest of that day and the next went by without any occurrence worthy of notice. Hanging by the feet to the front of the wire cover, on the side nearest to the light, the prisoner was motionless, inert. There was no oscillation of the wings, no tremor of the antennae, the female of the Great Peacock behaved in a similar fashion.

The female Bombyx gradually matured, her tender tissues gradually becoming firmer. By some process of which our scientists have not the least idea she elaborated a mysterious lure which would bring her lovers from the four corners of the sky. What was happening in this big-bellied body; what transmutations were accomplished, thus to affect the whole countryside?

On the third day the bride was ready. The festival opened brilliantly. I was in the garden, already despairing of success, for the days were passing and nothing had occurred, when towards three in the afternoon, the weather being very hot and the sun radiant, I perceived a crowd of butterflies gyrating in the embrasure of the open window.

The lovers had at last come to visit their lady. Some were emerging from the room, others were entering it; others, clinging to the wall of the house, were resting as though exhausted by a long journey. I could see others approaching in the distance, flying over the walls, over the screens of cypress. They came from all directions, but at last with decreasing frequency. I had missed the opening of the convocation, and now the gathering was almost complete.

I went indoors and upstairs. This time, in full daylight and without losing a detail, I witnessed once more the astonishing spectacle to which the great nocturnal butterfly had first introduced me. The study contained a cloud of males, which I estimated, at a glance, as being about sixty in number, so far as the movement and confusion allowed me to count them at all. After circling a few times over the cage many of them went to the open window, but returned immediately to recommence their evolutions. The most eager alighted on the cover, trampling on one another, jostling one another, trying to get the best places. On the other side of the barrier the captive, her great body hanging against the wire, waited immovable. She betrayed not a sign of emotion in the face of this turbulent swarm.

Going and entering, perched on the cover or fluttering round the room, for more than three hours they continued their frenzied saraband. But the sun was sinking, and the temperature was slowly falling. The ardour of the butterflies also cooled. Many went out not to return. Others took up their positions to wait for the gaieties of the following day; they clung to the cross-bars of the closed window as the males of the Great Peacock had done. The rejoicings were over for the day. They would certainly be renewed on the morrow, since the courtship was without result on account of the barrier of the wire-gauze cover.

But, alas I to my great disappointment, they were not resumed, and the fault was mine. Late in the day a Praying Mantis was brought to me, which merited attention on account of its exceptionally small size. Preoccupied with the events of the afternoon, and absent-minded, I hastily placed the predatory insect under the same cover as the moth. It did not occur to me for a moment that this cohabitation could lead to any harm. The Mantis was so slender, and the other so corpulent!

Alas! I little knew the fury of carnage animating the creature that wielded those tiny grappling-irons! Next morning I met with a disagreeable surprise: I found the little Mantis devouring the great moth. The head and the fore part of the thorax had already disappeared. Horrible creature! at what an evil hour you came to me! Goodbye to my researches, the plans which I had caressed all night in my imagination! For three years for lack of a subject, I was unable to resume them.

Bad luck, however, was not to make me forget the little I had learned. On one single occasion about sixty males had arrived. Considering the rarity of the Oak Eggar, and remembering the years of fruitless search on the part of my helpers and myself, this number was no less than stupefying. The undiscoverable had suddenly become multitudinous at the call of the female.

Whence did they come? From all sides, and undoubtedly from considerable distances. During my prolonged searches every bush and thicket and heap of stones in my neighbourhood had become familiar to me, and I can assert that the Oak Eggar was not to be found there. For such a swarm to collect as I found in my laboratory the moths must have come from all directions, from the whole district, and within a radius that I dare not guess at.

Three years went by and by chance two more cocoons of the Monk or Oak Eggar again fell into my hands. Both produced females, at an interval of a few days towards the middle of August; so that I was able to vary and repeat my experiments.

I rapidly repeated the experiments which had given me such positive results in the instance of the Great Peacock moth. The pilgrims of the day were no less skilful at finding their mates than the pilgrims of the night. They laughed at all my tricks. Infallibly they found the prisoners in their wire-gauze prisons, no matter in what part of the house they were placed; they discovered them in the depths of a wall-cupboard; they divined the secret of all manner of boxes, provided these were not rigorously air-tight. They came no longer when the box was hermetically sealed. So far this was only a repetition of the feats of the Great Peacock.

A box perfectly closed, so that the air contained therein had no communication with the external atmosphere, left the male in complete ignorance of the recluse. Not a single one arrived, even when the box was exposed and plain to see on the window-sill. Thus the idea of strongly scented effluvia, which are cut off by screens of wood, metal, card, glass, or what not, returns with double force.

I have shown that the great nocturnal moth was not thrown off the scent by the powerful odour of naphthaline, which I thought would mask the extra-subtle emanations of the female, which were imperceptible to human olfactory organs. I repeated the experiment with the Oak Eggar. This time I used all the resources of scent and stench that my knowledge of drugs would permit.

A dozen saucers were arranged, some in the interior of the wire-gauze cover, the prison of the female, and some around it, in an unbroken circle. Some contained naphthaline; others the essential oil of spike-lavender; others petroleum, and others a solution of alkaline sulphur giving off a stench of rotten eggs. Short of asphyxiating the prisoner I could do no more. These arrangements were made in the morning, so that the room should be saturated when the congregation of lovers should arrive.

In the afternoon the laboratory was filled with the most abominable stench, in which the penetrating aroma of spike-lavender and the stink of sulphuretted hydrogen were predominant. I must add that tobacco was habitually smoked in this room, and in abundance. The concerted odours of a gas-works, a smoking-room, a perfumery, a petroleum well, and a chemical factory--would they succeed in confusing the male moths?

By no means. About three o'clock the moths arrived in as great numbers as usual. They went straight to the cage, which I had covered with a thick cloth in order to add to their difficulties. Seeing nothing when once they had entered, and im-

mersed in an extraordinary atmosphere in which any subtle fragrance should have
been annihilated, they nevertheless made straight for the prisoner, and attempted
to reach her by burrowing under the linen cloth. My artifice had no result.

After this set-back, so obvious in its consequences, which only repeated the
lesson of the experiments made with naphthaline when my subject was the Great
Peacock, I ought logically to have abandoned the theory that the moths are guided
to their wedding festivities by means of strongly scented effluvia. That I did not do
so was due to a fortuitous observation. Chance often has a surprise in store which
sets us on the right road when we have been seeking it in vain.

One afternoon, while trying to determine whether sight plays any part in the
search for the female once the males had entered the room, I placed the female in
a bell-glass and gave her a slender twig of oak with withered leaves as a support.
The glass was set upon a table facing the open window. Upon entering the room the
moths could not fail to see the prisoner, as she stood directly in the way. The tray,
containing a layer of sand, on which the female had passed the preceding day and
night, covered with a wire-gauze dish-cover, was in my way. Without premedita-
tion I placed it at the other end of the room on the floor, in a corner where there
was but little light. It was a dozen yards away from the window.

The result of these preparations entirely upset my preconceived ideas. None of
the arrivals stopped at the bell-glass, where the female was plainly to be seen, the
light falling full upon her prison. Not a glance, not an inquiry. They all flew to the
further end of the room, into the dark corner where I had placed the tray and the
empty dish-cover.

They alighted on the wire dome, explored it persistently, beating their wings
and jostling one another. All the afternoon, until sunset, the moths danced about
the empty cage the same saraband that the actual presence of the female had previously evoked. Finally they departed: not all, for there were some that would not go,
held by some magical attractive force.

Truly a strange result! The moths collected where there was apparently nothing to attract them, and remained there, unpersuaded by the sense of sight; they
passed the bell-glass actually containing the female without halting for a moment,
although she must have been seen by many of the moths both going and coming.
Maddened by a lure, they paid no attention to the reality.

What was the lure that so deceived them? All the preceding night and all the morning the female had remained under the wire-gauze cover; sometimes clinging to the wire-work, sometimes resting on the sand in the tray. Whatever she touched--above all, apparently, with her distended abdomen--was impregnated, as a result of long contact, with a certain emanation. This was her lure, her love-philtre; this it was that revolutionised the Oak Eggar world. The sand retained it for some time and diffused the effluvium in turn.

They passed by the glass prison in which the female was then confined and hastened to the meshes of wire and the sand on which the magic philtre had been poured; they crowded round the deserted chamber where nothing of the magician remained but the odorous testimony of her sojourn.

The irresistible philtre requires time for its elaboration. I conceive of it as an exhalation which is given off during courtship and gradually saturates whatever is in contact with the motionless body of the female. If the bell-glass was placed directly on the table, or, still better, on a square of glass, the communication between the inside and the outside was insufficient, and the males, perceiving no odour, did not arrive so long as that condition of things obtained. It was plain that this failure of transmission was not due to the action of the glass as a screen simply, for if I established a free communication between the interior of the bell-glass and the open air by supporting it on three small blocks, the moths did not collect round it at once, although there were plenty in the room; but in the course of half an hour or so the feminine alembic began to operate, and the visitors crowded round the bell-glass as usual.

In possession of these data and this unexpected enlightenment I varied the experiments, but all pointed to the same conclusion. In the morning I established the female under the usual wire-gauze cover. For support I gave her a little twig of oak as before. There, motionless as if dead, she crouched for hours, half buried in the dry leaves, which would thus become impregnated with her emanations.

When the hour of the daily visits drew near I removed the twig, which was by then thoroughly saturated with the emanations, and laid it on a chair not far from the open window. On the other hand I left the female under the cover, plainly exposed on the table in the middle of the room.

The moths arrived as usual: first one, then two, then three, and presently five and six. They entered, flew out again, re-entered, mounted, descended, came and went, always in the neighbourhood of the window, not far from which was the chair on which the twig lay. None made for the large table, on which, a few steps further from the window, the female awaited them in the wire-gauze cover. They hesitated, that was plain; they were still seeking.

Finally they found. And what did they find? Simply the twig, which that morning had served the ample matron as bed. Their wings rapidly fluttering, they alighted on the foliage; they explored it over and under, probed it, raised it, and displaced it so that the twig finally fell to the floor. None the less they continued to probe between the leaves. Under the buffets and the draught of their wings and the clutches of their eager feet the little bundle of leaves ran along the floor like a scrap of paper patted by the paws of a cat.

While the twig was sliding away with its band of investigators two new arrivals appeared. The chair lay in their path. They stopped at it and searched eagerly at the very spot on which the twig had been lying. But with these, as with the others, the real object of their desires was there, close by, under a wire cover which was not even veiled. None took any note of it. On the floor, a handful of butterflies were still hustling the bunch of leaves on which the female had reposed that morning; others, on the chair, were still examining the spot where the twig had lain. The sun sank, and the hour of departure struck. Moreover, the emanations were growing feebler, were evaporating. Without more ado the visitors left. We bade them goodbye till the morrow.

The following tests showed me that the leaf-covered twig which accidentally enlightened me might be replaced by any other substance. Some time before the visitors were expected I placed the female on a bed of cloth or flannel, card or paper. I even subjected her to the rigours of a camp-bed of wood, glass, marble, and metal. All these objects, after a contact of sufficient duration, had the same attraction for the males as the female moth herself. They retained this property for a longer or shorter time, according to their nature. Cardboard, flannel, dust, sand, and porous objects retained it longest. Metals, marble, and glass, on the contrary, quickly lost their efficacy. Finally, anything on which the female had rested communicated its virtues by contact; witness the butterflies crowding on the straw-bottomed chair after the twig fell to the ground.

Using one of the most favourable materials--flannel, for example--I witnessed a curious sight. I placed a morsel of flannel on which the mother moth had been lying all the morning at the bottom of a long test-tube or narrow-necked bottle, just permitting of the passage of a male moth. The visitors entered the vessels, struggled, and did not know how to extricate themselves. I had devised a trap by means of which I could exterminate the tribe. Delivering the prisoners, and removing the flannel, which I placed in a perfectly closed box, I found that they re-entered the trap; attracted by the effluvia that the flannel had communicated to the glass.

I was now convinced. To call the moths of the countryside to the wedding-feast, to warn them at a distance and to guide them the nubile female emits an odour of extreme subtlety, imperceptible to our own olfactory sense-organs. Even with their noses touching the moth, none of my household has been able to perceive the faintest odour; not even the youngest, whose sensibility is as yet unvitiated.

This scent readily impregnates any object on which the female rests for any length of time, when this object becomes a centre of attraction as active as the moth herself until the effluvium is evaporated.

Nothing visible betrays the lure. On a sheet of paper, a recent resting-place, around which the visitors had crowded, there was no visible trace, no moisture; the surface was as clean as before the impregnation.

The product is elaborated slowly, and must accumulate a little before it reveals its full power. Taken from her couch and placed elsewhere the female loses her attractiveness for the moment and is an object of indifference; it is to the resting-place, saturated by long contact, that the arrivals fly. But the female soon regains her power.

The emission of the warning effluvium is more or less delayed according to the species. The recently metamorphosed female must mature a little and her organs must settle to their work. Born in the morning, the female of the Great Peacock moth sometimes has visitors the night of the same day; but more often on the second day, after a preparation of forty hours or so. The Oak Eggar does not publish her banns of marriage before the third or fourth day.

Let us return for a moment to the problematical function of the antennae. The male Oak Eggar has a sumptuous pair, as has the Great Peacock or Emperor Moth. Are we to regard these silky "feelers" as a kind of directing compass?--I resumed, but without attaching much importance to the matter, my previous experiment of amputation. None of those operated on returned. Do not let us draw conclusions from that fact alone. We saw in the case of the Great Peacock that more serious reasons than the truncation of the antennae made return as a rule impossible.

Moreover, a second Bombyx or Eggar, the Clover Moth, very like the Oak Eggar, and like it superbly plumed, poses us a very difficult problem. It is fairly abundant around my home; even in the orchard I find its cocoon, which is easily confounded with that of the Oak Eggar. I was at first deceived by the resemblance. From six cocoons, which I expected to yield Oak Eggars, I obtained, about the end of August, six females of the other species. Well: about these six females, born in my house, never a male appeared, although they were undoubtedly present in the neighbourhood.

If the ample and feathery antennae are truly sense-organs, which receive information of distant objects, why were not my richly plumed neighbours aware of what was passing in my study? Why did their feathery "feelers" leave them in ignorance of events which would have brought flocks of the other Eggar? Once more, the organ does not determine the aptitude. One individual or species is gifted, but another is not, despite an organic equality.

CHAPTER XVI
A TRUFFLE-HUNTER:
THE *BOLBOCERAS GALLICUS*

In the matter of physics we hear of nothing to-day but the Roentgen rays, which penetrate opaque bodies and photograph the invisible. A splendid discovery; but nothing very remarkable as compared with the surprises reserved for us by the future, when, better instructed as to the why and wherefore of things than now, and supplementing our feeble senses by means of science, we shall succeed in rivalling, however imperfectly, the sensorial acuteness of the lower animals.

How enviable, in how many cases, is the superiority of the beasts! It makes us realise the insufficiency of our impressions, and the very indifferent efficacy of our sense-organs; it proclaims realities which amaze us, so far are they beyond our own attributes.

A miserable caterpillar, the Processional caterpillar, found on the pine-tree, has its back covered with meteorological spiracles which sense the coming weather and foretell the storm; the bird of prey, that incomparable watchman, sees the fallen mule from the heights of the clouds; the blind bats guided their flight without collision through the inextricable labyrinth of threads devised by Spallanzani; the carrier pigeon, at a hundred leagues from home, infallibly regains its loft across immensities which it has never known; and within the limits of its more modest powers a bee, the Chalicodoma, also adventures into the unknown, accomplishing its long journey and returning to its group of cells.

Those who have never seen a dog seeking truffles have missed one of the finest achievements of the olfactory sense. Absorbed in his duties, the animal goes forward, scenting the wind, at a moderate pace. He stops, questions the soil with his nostrils, and, without excitement, scratches the earth a few times with one paw. "There it is, master!" his eyes seem to say: "there it is! On the faith of a dog, there are truffles here!"

He says truly. The master digs at the point indicated. If the spade goes astray the dog corrects the digger, sniffing at the bottom of the hole. Have no fear that stones and roots will confuse him; in spite of depth and obstacles, the truffle will be found. A dog's nose cannot lie.

I have referred to the dog's speciality as a subtle sense of smell. That is certainly what I mean, if you will understand by that that the nasal passages of the animal are the seat of the perceptive organ; but is the thing perceived always a simple smell in the vulgar acceptation of the term--an effluvium such as our own senses perceive? I have certain reasons for doubting this, which I will proceed to relate.

On various occasions I have had the good fortune to accompany a truffle-dog of first-class capacities on his rounds. Certainly there was not much outside show about him, this artist that I so desired to see at work; a dog of doubtful breed, placid and meditative; uncouth, ungroomed, and quite inadmissible to the intimacies of the hearthrug. Talent and poverty are often mated.

His master, a celebrated *rabassier*[5] of the village, being convinced that my object was not to steal his professional secrets, and so sooner or later to set up in business as a competitor, admitted me of his company, a favour of which he was not prodigal. From the moment of his regarding me not as an apprentice, but merely as a curious spectator, who drew and wrote about subterranean vegetable affairs, but had no wish to carry to market my bagful of these glories of the Christmas goose, the excellent man lent himself generously to my designs.

It was agreed between us that the dog should act according to his own instincts, receiving the customary reward, after each discovery, no matter what its size, of a crust of bread the size of a finger-nail. Every spot scratched by his paw should be excavated, and the object indicated was to be extracted without reference to its marketable value. In no case was the experience of the master to intervene in order to divert the dog from a spot where the general aspect of things indicated that no commercial results need be expected, for I was more concerned with the miserable specimens unfit for the market than with the choice specimens, though of course the latter were welcomed.

Thus conducted, this subterranean botanising was extremely fruitful. With that perspicacious nose of his the dog obtained for me both large and small, fresh and putrid, odorous and inodorous, fragrant and offensive. I was amazed at my collection, which comprised the greater number of the hypogenous fungi of the neighbourhood.

What a variety of structure, and above all of odour, the primordial quality in this question of scent! There were some that had no appreciable scent beyond a vague fungoid flavour, more or less common to all. Others smelt of turnips, of sour cabbage; some were fetid, sufficiently so to make the house of the collector noisome. Only the true truffle possessed the aroma dear to epicures. If odour, as we understand it, is the dog's only guide, how does he manage to follow that guide amidst all these totally different odours? Is he warned of the contents of the subsoil by a general emanation, by that fungoid effluvium common to all the species? Thus a somewhat embarrassing question arises.

I paid special attention to the ordinary toadstools and mushrooms, which announced their near advent by cracking the surface of the soil. Now these points, where my eyes divined the cryptogam pushing back the soil with its button-like heads, these points, where the ordinary fungoid odour was certainly very pronounced, were never selected by the dog. He passed them disdainfully, without a sniff, without a stroke of the paw. Yet the fungi were underground, and their odour was similar to that I have already referred to.

I came back from my outings with the conviction that the truffle-finding nose has some better guide than odour such as we with our sense-organs conceive it. It must perceive effluvia of another order as well; entirely mysterious to us, and therefore not utilised. Light has its dark rays--rays without effect upon our retinas, but not apparently on all. Why should not the domain of smell have its secret emanations, unknown to our senses and perceptible to a different sense-organ?

If the scent of the dog leaves us perplexed in the sense that we cannot possibly say precisely, cannot even suspect what it is that the dog perceives, at least it is clear that it would be erroneous to refer everything to human standards. The world of sensations is far larger than the limits of our own sensibility. What numbers of facts relating to the interplay of natural forces must escape us for want of sufficiently sensitive organs!

The unknown--that inexhaustible field in which the men of the future will try their strength--has harvests in store for us beside which our present knowledge would show as no more than a wretched gleaning. Under the sickle of science will one day fall the sheaves whose grain would appear to-day as senseless paradoxes. Scientific dreams? No, if you please, but undeniable positive realities, affirmed by the brute creation, which in certain respects has so great an advantage over us.

Despite his long practice of his calling, despite the scent of the object he was seeking, the *rabassier* could not divine the presence of the truffle, which ripens in winter under the soil, at a depth of a foot or two; he must have the help of a dog or a pig, whose scent is able to discover the secrets of the soil. These secrets are known to various insects even better than to our two auxiliaries. They have in exceptional perfection the power of discovering the tubers on which their larvae are nourished.

From truffles dug up in a spoiled condition, peopled with vermin, and placed in that condition, with a bed of fresh sand, in a glass jar, I have in the past obtained a small red beetle, known as the truffle-beetle (Anisotoma cinnamomea, Panz.), and various Diptera, among which is a Sapromyzon which, by its sluggish flight and its fragile form, recalls the *Scatophaga scybalaria*, the yellow velvety fly which is found in human excrement in the autumn. The latter finds its refuge on the surface of the soil, at the foot of a wall or hedge or under a bush; but how does the former know just where the truffle lies under the soil, or at what depth? To penetrate to that depth, or to seek in the subsoil, is impossible. Its fragile limbs, barely able to move a grain of sand, its extended wings, which would bar all progress in a narrow passage, and its costume of bristling silken pile, which would prevent it from slipping through crevices, all make such a task impossible. The Sapromyzon is forced to lay its eggs on the surface of the soil, but it does so on the precise spot which overlies the truffle, for the grubs would perish if they had to wander at random in search of their provender, the truffle being always thinly sown.

The truffle fly is informed by the sense of smell of the points favourable to its maternal plans; it has the talents of the truffle-dog, and doubtless in a higher degree, for it knows naturally, without having been taught, what its rival only acquires through an artificial education.

It would be not uninteresting to follow the Sapromyzon in its search in the open woods. Such a feat did not strike me as particularly possible; the insect is rare, flies off quickly when alarmed, and is lost to view. To observe it closely under such conditions would mean a loss of time and an assiduity of which I do not feel capable. Another truffle-hunter will show us what we could hardly learn from the fly.

This is a pretty little black beetle, with a pale, velvety abdomen; a spherical insect, as large as a biggish cherry-stone. Its official title is ***Bolboceras gallicus***, Muls. By rubbing the end of the abdomen against the edge of the wing-cases it produces a gentle chirping sound like the cheeping of nestlings when the mother-bird returns to the nest with food. The male wears a graceful horn on his head; a duplicate, in little, of that of the ***Copris hispanus***.

Deceived by this horn, I at first took the insect for a member of the corporation of dung-beetles, and as such I reared it in captivity. I offered it the kind of diet most appreciated by its supposed relatives, but never, never would it touch such food. For whom did I take it? Fie upon me! To offer ordure to an epicure! It required, if not precisely the truffle known to our ***chefs*** and ***gourmets***, at least its equivalent.

This characteristic I grasped only after patient investigation. At the southern foot of the hills of Serignan, not far from the village, is a wood of maritime pines alternating with rows of cypress. There, towards Toussaint, after the autumnal rains, you may find an abundance of the mushrooms or "toadstools" that affect the conifers; especially the delicious Lactaris, which turns green if the points are rubbed and drips blood if broken. In the warm days of autumn this is the favourite promenade of the members of my household, being distant enough to exercise their young legs, but near enough not to fatigue them.

There one finds and sees all manner of things: old magpies' nests, great bundles of twigs; jays, wrangling after filling their crops with the acorns of the neighbouring oaks; rabbits, whose little white upturned scuts go bobbing away through the rosemary bushes; dung-beetles, which are storing food for the winter and throwing up their rubbish on the threshold of their burrows. And then the fine sand, soft to the touch, easily tunnelled, easily excavated or built into tiny huts which we thatch with moss and surmount with the end of a reed for a chimney; and the delicious meal of apples, and the sound of the aeolian harps which softly whisper among the boughs of the pines!

For the children it is a real paradise, where they can receive the reward of well-learned lessons. The grown-ups also can share in the enjoyment. As for myself, for long years I have watched two insects which are found there without getting to the bottom of their domestic secrets. One is the ***Minotaurus typhaeus***, whose male carries on his corselet three spines which point forward. The old writers called him the Phalangist, on account of his armour, which is comparable to the three ranks of lances of the Macedonian phalanx.

This is a robust creature, heedless of the winter. All during the cold season, whenever the weather relents a little, it issues discreetly from its lodging, at nightfall, and gathers, in the immediate neighbourhood of its dwelling, a few fragments of sheep-dung and ancient olives which the summer suns have dried. It stacks them in a row at the end of its burrow, closes the door, and consumes them. When the food is broken up and exhausted of its meagre juices it returns to the surface and renews its store. Thus the winter passes, famine being unknown unless the weather is exceptionally hard.

The second insect which I have observed for so long among the pines is the Bolboceras. Its burrows, scattered here and there, higgledy-piggledy with those of the Minotaur, are easy to recognise. The burrow of the Phalangist is surmounted by a voluminous rubbish-dump, the materials of which are piled in the form of a cylinder as long as the finger. Each of these dumps is a load of refuse and rubbish pushed outward by the little sapper, which shoulders it up from below. The orifice is closed whenever the insect is at home, enlarging its tunnel or peacefully enjoying the contents of its larder.

The lodging of the Bolboceras is open and surrounded simply by a mound of sand. Its depth is not great; a foot or hardly more. It descends vertically in an easily shifted soil. It is therefore easy to inspect it, if we take care first of all to dig a trench so that the wall of the burrow may be afterwards cut away, slice by slice, with the blade of a knife. The burrow is thus laid bare along its whole extent, from the surface to the bottom, until nothing remains of it but a demi-cylindrical groove.

Often the violated dwelling is empty. The insect has departed in the night, having finished its business there. It is a nomad, a night-walker, which leaves its dwelling without regret and easily acquires another. Often, on the other hand, the insect will be found at the bottom of the burrow; sometimes a male, sometimes a female, but always alone. The two sexes, equally zealous in excavating their burrows, work apart without collaboration. This is no family mansion for the rearing of offspring; it is a temporary dwelling, made by each insect for its own benefit.

Sometimes the burrow contains nothing but the well-sinker surprised at its work: sometimes--and not rarely--the hermit will be found embracing a small subterranean fungus, entire or partly consumed. It presses it convulsively to its bosom and will not be parted from it. This is the insect's booty: its worldly wealth. Scattered crumbs inform us that we have surprised the beetle at a feast.

Let us deprive the insect of its booty. We find a sort of irregular, rugged, purse-like object, varying in size from the largeness of a pea to that of a cherry. The exterior is reddish, covered with fine warts, having an appearance not unlike shagreen; the interior, which has no communication with the exterior, is smooth and white. The pores, ovoidal and diaphanous, are contained, in groups of eight, in long capsules. From these characteristics we recognise an underground cryptogam, known to the botanists as *Hydnocystis arenaria*, and a relation of the truffle.

This discovery begins to throw a light on the habits of the Bolboceras and the cause of its burrows, so frequently renewed. In the calm of the twilight the little truffle-hunter goes abroad, chirping softly to encourage itself. It explores the soil, and interrogates it as to its contents, exactly as does the truffle-gatherer's dog. The sense of smell warns it that the desired object is beneath it, covered by a few inches of sand. Certain of the precise point where the treasure lies, it sinks a well vertically downwards, and infallibly reaches it. So long as there is food left it does not again leave the burrow. It feasts happily at the bottom of its well, heedless of the open or imperfectly closed burrow.

When no more food is left it removes in search of further booty, which becomes the occasion of another burrow, this too in its turn to be abandoned. So many truffles eaten necessitate so many burrows, which are mere dining-rooms or pilgrim's larders. Thus pass the autumn and the spring, the seasons of the *Hydnocystis*, in the pleasures of the table and removal from one house to another.

To study the insect *rabassier* in my own house I had to obtain a small store of its favourite food. To seek it myself, by digging at random, would have resulted merely in waste of time; the little cryptogam is not so common that I could hope to find it without a guide. The truffle-hunter must have his dog; my guide should be the Bolboceras itself. Behold me, then, a ***rabassier*** of a kind hitherto unknown. I have told my secret, although I fear my original teacher will laugh at me if he ever hears of my singular form of competition.

The subterranean fungi grow only at certain points, but they are often found in groups. Now, the beetle has passed this way; with its subtle sense of smell it has recognised the ground as favourable; for its burrows are numerous. Let us dig, then, in the neighbourhood of these holes. The sign is reliable; in a few hours, thanks to the signs of the Bolboceras, I obtain a handful of specimens of the ***Hydnocystis***. It is the first time I have ever found this fungus in the ground. Let us now capture the insect--an easy matter, for we have only to excavate the burrows.

The same evening I begin my experiments. A wide earthen pan is filled with fresh sand which has been passed through a sieve. With the aid of a stick the thickness of a finger I make six vertical holes in the sand: they are conveniently far apart, and are eight inches in depth. A ***Hydnocystis*** is placed at the bottom of each; a fine straw is then inserted, to show me the precise position later. Finally the six holes are filled with sand which is beaten down so that all is firm. When the surface is perfectly level, and everywhere the same, except for the six straws, which mean nothing to the insect, I release my beetles, covering them with a wire-gauze cover. They are eight in number.

At first I see nothing but the inevitable fatigue due to the incidents of exhumation, transport, and confinement in a strange place. My exiles try to escape: they climb the wire walls, and finally all take to earth at the edge of their enclosure. Night comes, and all is quiet. Two hours later I pay my prisoners a last visit. Three are still buried under a thin layer of sand. The other five have sunk each a vertical well at the very foot of the straws which indicate the position of the buried fungi. Next morning the sixth straw has its burrow like the rest.

It is time to see what is happening underground. The sand is methodically removed in vertical slices. At the bottom of each burrow is a Bolboceras engaged in eating its truffle.

Let us repeat the experiment with the partly eaten fungi. The result is the same. In one short night the food is divined under its covering of sand and attained by means of a burrow which descends as straight as a plumb-line to the point where the fungus lies. There has been no hesitation, no trial excavations which have nearly discovered the object of search. This is proved by the surface of the soil, which is everywhere just as I left it when smoothing it down. The insect could not make more directly for the objective if guided by the sense of sight; it digs always at the foot of the straw, my private sign. The truffle-dog, sniffing the ground in search of truffles, hardly attains this degree of precision.

Does the **Hydnocystis** possess a very keen odour, such as we should expect to give an unmistakable warning to the senses of the consumer? By no means. To our own sense of smell it is a neutral sort of object, with no appreciable scent whatever. A little pebble taken from the soil would affect our senses quite as strongly with its vague savour of fresh earth. As a finder of underground fungi the Bolboceras is the rival of the dog. It would be the superior of the dog if it could generalise; it is, however, a rigid specialist, recognising nothing but the **Hydnocystis**. No other fungus, to my knowledge, either attracts it or induces it to dig.[6]

Both dog and beetle are very near the subsoil which they scrutinise; the object they seek is at no great depth. At a greater depth neither dog nor insect could perceive such subtle effluvia, nor even the odour of the truffle. To attract insect or animal at a great distance powerful odours are necessary, such as our grosser senses can perceive. Then the exploiters of the odorous substance hasten from afar off and from all directions.

If for purposes of study I require specimens of such insects as dissect dead bodies I expose a dead mole to the sunlight in a distant corner of my orchard. As soon as the creature is swollen with the gases of putrefaction, and the fur commences to fall from the greenish skin, a host of insects arrive--Silphidae, Dermestes, Horn-beetles, and Necrophori--of which not a single specimen could ever be obtained in my garden or even in the neighbourhood without the use of such a bait.

They have been warned by the sense of smell, although far away in all directions, while I myself can escape from the stench by recoiling a few paces. In comparison with their sense of smell mine is miserable; but in this case, both for me and for them, there is really what our language calls an odour.

I can do still better with the flower of the Serpent Arum (Arum dracunculus), so noteworthy both for its form and its incomparable stench. Imagine a wide lanceolated blade of a vinous purple, some twenty inches in length, which is twisted at the base into an ovoid purse about the size of a hen's egg. Through the opening of this capsule rises the central column, a long club of a livid green, surrounded at the base by two rings, one of ovaries and the other of stamens. Such, briefly, is the flower or rather the inflorescence of the Serpent Arum.

For two days it exhales a horrible stench of putrid flesh; a dead dog could not produce such a terrible odour. Set free by the sun and the wind, it is odious, intolerable. Let us brave the infected atmosphere and approach; we shall witness a curious spectacle.

Warned by the stench, which travels far and wide, a host of insects are flying hither; such insects as dissect the corpses of frogs, adders, lizards, hedgehogs, moles and field-mice--creatures that the peasant finds beneath his spade and throws disembowelled on the path. They fall upon the great leaf, whose livid purple gives it the appearance of a strip of putrid flesh; they dance with impatience, intoxicated by the corpse-like odour which to them is so delicious; they roll down its steep face and are engulfed in the capsule. After a few hours of hot sunlight the receptacle is full. Let us look into the capsule through the narrow opening. Nowhere else could you see such a mob of insects. It is a delirious mixture of backs and bellies, wing-covers and legs, which swarms and rolls upon itself, rising and falling, seething and boiling, shaken by continual convulsions, clicking and squeaking with a sound of entangled articulations. It is a bacchanal, a general access of delirium tremens.

A few, but only a few, emerge from the mass. By the central mast or the walls of the purse they climb to the opening. Do they wish to take flight and escape? By no means. On the threshold of the cavity, while already almost at liberty, they allow themselves to fall into the whirlpool, retaken by their madness. The lure is irresistible. None will break free from the swarm until the evening, or perhaps the next day, when the heady fumes will have evaporated. Then the units of the swarm disengage themselves from their mutual embraces, and slowly, as though regretfully, take flight and depart. At the bottom of this devil's purse remains a heap of the dead and dying, of severed limbs and wing-covers torn off; the inevitable sequels of the frantic orgy. Soon the woodlice, earwigs, and ants will appear to prey upon the injured.

What are these insects doing? Were they the prisoners of the flower, converted into a trap which allowed them to enter but prevented their escape by means of a palisade of converging hairs? No, they were not prisoners; they had full liberty to escape, as is proved by the final exodus, which is in no way impeded. Deceived by a fallacious odour, were they endeavouring to lay and establish their eggs as they would have done under the shelter of a corpse? No; there is no trace of eggs in the purse of the Arum. They came convoked by the odour of a decaying body, their supreme delight; an intoxication seized them, and they rushed into the eddying swarm to take part in a festival of carrion-eaters.

I was anxious to count the number of those attracted. At the height of the bacchanal I emptied the purse into a bottle. Intoxicated as they were, many would escape my census, and I wished to ensure its accuracy. A few drops of carbon bisulphide quieted the swarm. The census proved that there were more than four hundred insects in the purse of the Arum. The collection consisted entirely of two species--Dermestes and Saprinidae--both eager prospectors of carrion and animal detritus during the spring.

My friend Bull, an honest dog all his lifetime if ever there was one, amongst other eccentricities had the following: finding in the dust of the road the shrivelled body of a mole, flattened by the feet of pedestrians, mummified by the heat of the sun, he would slide himself over it, from the tip of his nose to the root of his tail, he would rub himself against it deliciously over and over again, shaken with nervous spasms, and roll upon it first in one direction, then in the other.

It was his sachet of musk, his flask of eau-de-Cologne. Perfumed to his liking, he would rise, shake himself, and proceed on his way, delighted with his toilet. Do not let us scold him, and above all do not let us discuss the matter. There are all kinds of tastes in a world.

Why should there not be insects with similar habits among the amateurs of corpse-like savours? We see Dermestes and Saprinidae hastening to the arum-flower. All day long they writhe and wriggle in a swarm, although perfectly free to escape; numbers perish in the tumultuous orgy. They are not retained by the desire of food, for the arum provides them with nothing eatable; they do not come to breed, for they take care not to establish their grubs in that place of famine. What are these frenzied creatures doing? Apparently they are intoxicated with fetidity, as was Bull when he rolled on the putrid body of a mole.

This intoxication draws them from all parts of the neighbourhood, perhaps over considerable distances; how far we do not know. The Necrophori, in quest of a place where to establish their family, travel great distances to find the corpses of small animals, informed by such odours as offend our own senses at a considerable distance.

The **Hydnocystis**, the food of the Bolboceras, emits no such brutal emanations as these, which readily diffuse themselves through space; it is inodorous, at least to our senses. The insect which seeks it does not come from a distance; it inhabits the places wherein the cryptogam is found. Faint as are the effluvia of this subterranean fungus, the prospecting epicure, being specially equipped, perceives them with the greatest ease; but then he operates at close range, from the surface of the soil. The truffle-dog is in the same case; he searches with his nose to the ground. The true truffle, however, the essential object of his search, possesses a fairly vivid odour.

But what are we to say of the Great Peacock moth and the Oak Eggar, both of which find their captive female? They come from the confines of the horizon. What do they perceive at that distance? Is it really an odour such as we perceive and understand? I cannot bring myself to believe it.

The dog finds the truffle by smelling the earth quite close to the tuber; but he finds his master at great distances by following his footsteps, which he recognises by their scent. Yet can he find the truffle at a hundred yards? or his master, in the complete absence of a trail? No. With all his fineness of scent, the dog is incapable of such feats as are realised by the moth, which is embarrassed neither by distance nor the absence of a trail. It is admitted that odour, such as affects our olfactory sense, consists of molecules emanating from the body whose odour is perceived. The odorous material becomes diffused through the air to which it communicates its agreeable or disagreeable aroma. Odour and taste are to a certain extent the same; in both there is contact between the material particles causing the impression and the sensitive papillae affected by the impression.

That the Serpent Arum should elaborate a powerful essence which impregnates the atmosphere and makes it noisome is perfectly simple and comprehensible. Thus the Dermestes and Saprinidae, those lovers of corpse-like odours, are warned by molecular diffusion. In the same way the putrid frog emits and disseminates around it atoms of putrescence which travel to a considerable distance and so attract and delight the Necrophorus, the carrion-beetle.

But in the case of the Great Peacock or the Oak Eggar, what molecules are actually disengaged? None, according to our sense of smell. And yet this lure, to which the males hasten so speedily, must saturate with its molecules an enormous hemisphere of air--a hemisphere some miles in diameter! What the atrocious fetor of the Arum cannot do the absence of odour accomplishes! However divisible matter may be, the mind refuses such conclusions. It would be to redden a lake with a grain of carmine; to fill space with a mere nothing.

Moreover, where my laboratory was previously saturated with powerful odours which should have overcome and annihilated any particularly delicate effluvium, the male moths arrived without the least indication of confusion or delay.

A loud noise stifles a feeble note and prevents it from being heard; a brilliant light eclipses a feeble glimmer. Heavy waves overcome and obliterate ripples. In the two cases cited we have waves of the same nature. But a clap of thunder does not diminish the feeblest jet of light; the dazzling glory of the sun will not muffle the slightest sound. Of different natures, light and sound do not mutually interact.

My experiment with spike-lavender, naphthaline, and other odours seems to prove that odour proceeds from two sources. For emission substitute undulation, and the problem of the Great Peacock moth is explained. Without any material emanation a luminous point shakes the ether with its vibrations and fills with light a sphere of indefinite magnitude. So, or in some such manner, must the warning effluvium of the mother Oak Eggar operate. The moth does not emit molecules; but something about it vibrates, causing waves capable of propagation to distances incompatible with an actual diffusion of matter.

From this point of view, smell would have two domains--that of particles dissolved in the air and that of etheric waves.[7] The former domain alone is known to us. It is also known to the insect. It is this that warns the Saprinidae of the fetid arum, the Silphidae and the Necrophori of the putrid mole.

The second category of odour, far superior in its action through space, escapes us completely, because we lack the essential sensory equipment. The Great Peacock moth and the Oak Eggar know it at the time of their nuptial festivities. Many others must share it in differing degrees, according to the exigencies of their way of life.

Like light, odour has its X-rays. Let science, instructed by the insect, one day give us a radiograph sensitive to odours, and this artificial nose will open a new world of marvels.

CHAPTER XVII
THE ELEPHANT-BEETLE

Some of our machines have extraordinary-looking mechanisms, which remain inexplicable so long as they are seen in repose. But wait until the whole is in motion; then the uncouth-looking contrivance, with its cog-wheels interacting and its connecting-rods oscillating, will reveal the ingenious combination in which all things are skilfully disposed to produce the desired effects. It is the same with certain insects; with certain weevils, for instance, and notably with the Acorn-beetles or Balanini, which are adapted, as their name denotes, to the exploitation of acorns, nuts, and other similar fruits.

The most remarkable, in my part of France, is the Acorn Elephant (Balaninus elephas, Sch.). It is well named; the very name evokes a mental picture of the insect. It is a living caricature, this beetle with the prodigious snout. The latter is no thicker than a horsehair, reddish in colour, almost rectilinear, and of such length that in order not to stumble the insect is forced to carry it stiffly outstretched like a lance in rest. What is the use of this embarrassing pike, this ridiculous snout?

Here I can see some reader shrug his shoulders. Well, if the only end of life is to make money by hook or by crook, such questions are certainly ridiculous.

Happily there are some to whom nothing in the majestic riddle of the universe is little. They know of what humble materials the bread of thought is kneaded; a nutriment no less necessary than the bread made from wheat; and they know that both labourers and inquirers nourish the world with an accumulation of crumbs.

Let us take pity on the question, and proceed. Without seeing it at work, we already suspect that the fantastic beak of the Balaninus is a drill analogous to those which we ourselves use in order to perforate hard materials. Two diamond-points, the mandibles, form the terminal armature of the drill. Like the Larinidae, but under conditions of greater difficulty, the Curculionidae must use the implement in order to prepare the way for the installation of their eggs.

But however well founded our suspicion may be, it is not a certitude. I can only discover the secret by watching the insect at work.

Chance, the servant of those that patiently solicit it, grants me a sight of the acorn-beetle at work, in the earlier half of October. My surprise is great, for at this late season all industrial activity is as a rule at an end. The first touch of cold and the entomological season is over.

To-day, moreover, it is wild weather; the *bise* is moaning, glacial, cracking one's lips. One needs a robust faith to go out on such a day in order to inspect the thickets. Yet if the beetle with the long beak exploits the acorns, as I think it does, the time presses if I am to catch it at its work. The acorns, still green, have acquired their full growth. In two or three weeks they will attain the chestnut brown of perfect maturity, quickly followed by their fall.

My seemingly futile pilgrimage ends in success. On the evergreen oaks I surprise a Balaninus with the trunk half sunk in an acorn. Careful observation is impossible while the branches are shaken by the *mistral*. I detach the twig and lay it gently upon the ground. The insect takes no notice of its removal; it continues its work. I crouch beside it, sheltered from the storm behind a mass of underwood, and watch operations.

Shod with adhesive sandals which later on, in my laboratory, will allow it rapidly to climb a vertical sheet of glass, the elephant-beetle is solidly established on the smooth, steep curvature of the acorn. It is working its drill. Slowly and awkwardly it moves around its implanted weapon, describing a semicircle whose centre is the point of the drill, and then another semicircle in the reverse direction. This is repeated over and over again; the movement, in short, is identical with that we give to a bradawl when boring a hole in a plank.

Little by little the rostrum sinks into the acorn. At the end of an hour it has entirely disappeared. A short period of repose follows, and finally the instrument is withdrawn. What is going to happen next? Nothing on this occasion. The Balaninus abandons its work and solemnly retires, disappearing among the withered leaves. For the day there is nothing more to be learned.

But my interest is now awakened. On calm days, more favourable to the entomologist, I return to the woods, and I soon have sufficient insects to people my laboratory cages. Foreseeing a serious difficulty in the slowness with which the beetle labours, I prefer to study them indoors, with the unlimited leisure only to be found in one's own home.

The precaution is fortunate. If I had tried to continue as I began, and to observe the Balaninus in the liberty of the woods, I should never, even with the greatest good fortune, have had the patience to follow to the end the choice of the acorn, the boring of the hole, and the laying of the eggs, so meticulously deliberate is the insect in all its affairs; as the reader will soon be able to judge.

Three species of oak-tree compose the copse inhabited by the Balaninus: the evergreen oak and the pubescent oak, which would become fine trees if the woodman would give them time, and the kermes oak, a mere scrubby bush. The first species, which is the most abundant of the three, is that preferred by the Balaninus. The acorn is firm, elongated, and of moderate size; the cup is covered with little warts. The acorns of the pubescent oak are usually stunted, short, wrinkled, and fluted, and subject to premature fall. The aridity of the hills of Serignan is unfavourable to them. The Acorn-beetles accept them only in default of something better.

The kermes, a dwarf oak, a ridiculous tree which a man can jump over, surprises me by the wealth of its acorns, which are large, ovoidal growths, the cup being covered with scales. The Balaninus could not make a better choice; the acorn affords a safe, strong dwelling and a capacious storehouse of food.

A few twigs from these three trees, well provided with acorns, are arranged under the domes of some of my wire-gauze covers, the ends being plunged into a glass of water which will keep them fresh. A suitable number of couples are then introduced into the cages; and the latter are placed at the windows of my study, where they obtain the direct sunlight for the greater part of the day. Let us now arm ourselves with patience, and keep a constant watch upon events. We shall be rewarded; the exploitation of the acorn deserves to be seen.

Matters do not drag on for very long. Two days after these preparations I arrive at the precise moment when the task is commenced. The mother, larger than the male, and equipped with a longer drill, is inspecting her acorn, doubtless with a view to depositing her eggs.

She goes over it step by step, from the point to the stem, both above and below. On the warty cup progression is easy; over the rest of the surface it would be impossible, were not the soles of her feet shod with adhesive pads, which enable her to retain her hold in any position. Without the least uncertainty of footing, the insect walks with equal facility over the top or bottom or up the sides of the slippery fruit.

The choice is made; the acorn is recognised as being of good quality. The time has come to sink the hole. On account of its excessive length it is not easy to manoeuvre the beak. To obtain the best mechanical effect the instrument must be applied perpendicularly to the convex surface of the acorn, and the embarrassing implement which is carried in front of the insect when the latter is not at work must now be held in such a position as to be beneath the worker.

To obtain this result the insect rears herself upon her hind legs, supporting herself upon the tripod formed by the end of the wing-covers and the posterior tarsi. It would be hard to imagine anything more curious than this little carpenter, as she stands upright and brings her nasal bradawl down towards her body.

Now the drill is held plumb against the surface, and the boring commences. The method is that I witnessed in the wood on the day of the storm. Very slowly the insect veers round from right to left, then from left to right. Her drill is not a spiral gimlet which will sink itself by a constant rotary motion; it is a bradawl, or rather a trochar, which progresses by little bites, by alternative erosion, first in one direction, then the other.

Before continuing, let me record an accident which is too striking to be passed over. On various occasions I have found the insect dead in the midst of its task. The body is in an extraordinary position, which would be laughable if death were not always a serious thing, above all when it comes suddenly, in the midst of labour.

The drill is implanted in the acorn just a little beyond the tip; the work was only commenced. At the top of the drill, at right angles to it, the Balaninus is suspended in the air, far from the supporting surface of the acorn. It is dried, mummified, dead I know not how long. The legs are rigid and contracted under the body. Even if they retained the flexibility and the power of extension that were theirs in life, they would fall far short of the surface of the acorn. What then has happened, that this unhappy insect should be impaled like a specimen beetle with a pin through its head?

An accident of the workshop is responsible. On account of the length of its implement the beetle commences her work standing upright, supported by the two hind-legs. Imagine a slip, a false step on the part of the two adhesive feet; the unfortunate creature will immediately lose her footing, dragged by the elasticity of the snout, which she was forced to bend somewhat at the beginning. Torn away from her foothold, the suspended insect vainly struggles in air; nowhere can her feet, those safety anchors, find a hold. She starves at the end of her snout, for lack of foothold whereby to extricate herself. Like the artisans in our factories, the elephant-beetle is sometimes the victim of her tools. Let us wish her good luck, and sure feet, careful not to slip, and proceed.

On this occasion all goes well, but so slowly that the descent of the drill, even when amplified by the magnifying-glass, cannot be perceived. The insect veers round perpetually, rests, and resumes her work. An hour passes, two hours, wearying the observer by their sustained attention; for I wish to witness the precise moment when the beetle withdraws her drill, turns round, and deposits her egg in the mouth of the orifice. This, at least, is how I foresee the event.

Two hours go by, exhausting my patience. I call the household to my aid. Three of us take turns, keeping an uninterrupted watch upon the persevering creature whose secret I intend at any cost to discover.

It was well that I called in helpers to lend me their eyes and their attention. After eight hours--eight interminable hours, when it was nearly night, the sentinel on the watch calls me. The insect appears to have finished. She does, in fact, very cautiously withdraw her beak, as though fearing to slip. Once the tool is withdrawn she holds it pointing directly in front of her.

The moment has come.... Alas, no! Once more I am cheated; my eight hours of observation have been fruitless. The Balaninus decamps; abandons her acorn without laying her eggs. I was certainly right to distrust the result of observation in the open woods. Such concentration among the oaks, exposed to the sun, wind, and rain would have been an intolerable task.

During the whole of October, with the aid of such helpers as are needed, I remark a number of borings, not followed by the laying of eggs. The duration of the observer's task varies greatly. It usually amounts to a couple of hours; sometimes it exceeds half the day.

With what object are these perforations made, so laborious and yet so often unused? Let us first of all discover the position of the egg, and the first mouthfuls taken by the grub, and perhaps the reply will be found.

The peopled acorns remain on the oak, held in their cups as though nothing had occurred to the detriment of the cotyledons. With a little attention they may be readily recognised. Not far from the cup, on the smooth, still green envelope of the acorn a little point is visible; a tiny needle-prick. A narrow brown aureole, the product of mortification, is not long in appearing. This marks the opening of the hole. Sometimes, but more rarely, the hole is drilled through the cup itself.

Let us select those acorns which have been recently perforated: that is to say, those in which the perforation is not yet surrounded by the brown ring which appears in course of time. Let us shell them. Many contain nothing out of the way; the Balaninus has bored them but has not laid her eggs in them. They resemble the acorns which for hours and hours were drilled in my laboratory but not utilised. Many, on the contrary, contain an egg.

Now however distant the entrance of the bore may be, this egg is always at the bottom of the acorn, within the cup, at the base of the cotyledonary matter. The cup furnishes a thin film like swan-skin which imbibes the sapid exudations from the stem, the source of nourishment. I have seen a young grub, hatched under my eyes, eat as his first mouthfuls this tender cottony layer, which is moist and flavoured with tannin.

Such nutriment, juicy and easy of digestion, like all nascent organic matter, is only found in this particular spot; and it is only there, between the cup and the base of the cotyledons, that the elephant-beetle establishes her egg. The insect knows to a nicety the position of the portions best adapted to the feeble stomach of the newly hatched larva.

Above this is the tougher nutriment of the cotyledons. Refreshed by its first meal, the grub proceeds to attack this; not directly, but in the tunnel bored by the mother, which is littered with tiny crumbs and half-masticated shavings. With this light mealy diet the strength of the grub increases, and it then plunges directly into the substance of the acorn.

These data explain the tactics of the gravid mother. What is her object when, before proceeding to sink her hole, she inspects her acorn, from above, below, before and behind, with such meticulous care? She is making sure that the acorn is not already occupied. The larder is amply stored, but it does not contain enough for two. Never in fact, have I found two larvae in the same acorn. One only, always only one, digests the copious meal and converts it into a greenish dust before leaving it and descending to the ground. Only an insignificant shell remains uneaten. The rule is, to each grub one acorn.

Before trusting the egg to the acorn it is therefore essential to subject it to a thorough examination, to discover whether it already has an occupant. This possible occupant would be at the base of the acorn, under the cover of the cup. Nothing could be more secret than this hiding-place. Not an eye could divine the inhabitant if the surface of the acorn did not bear the mark of a tiny perforation.

This mark, just visible, is my guide. Its presence tells me that the acorn is inhabited, or at least that it has been prepared for the reception of the egg; its absence tells me that the acorn has not yet been appropriated. The elephant-beetle undoubtedly draws the same conclusions.

I see matters from on high, with a comprehensive glance, assisted at will by the magnifying-glass. I turn the acorn between my fingers for a moment, and the inspection is concluded. The beetle, investigating the acorn at close quarters, is often obliged to scrutinise practically the entire surface before detecting the tell-tale spot. Moreover, the welfare of her family demands a far more careful search than does my curiosity. This is the reason for her prolonged and deliberate examination.

The search is concluded; the acorn is recognised as unoccupied. The drill is applied to the surface and rotated for hours; then, very often, the insect departs, disdaining the result of her work. Why such protracted efforts? Was the beetle piercing the fruit merely to obtain drink and refreshment? Was the beak thrust into the depths of the base merely to obtain, from the choicer parts, a few sips of nutritious sap? Was the whole undertaking merely a matter of personal nourishment?

At first I believed this to be the solution, though surprised at the display of so much perseverance rewarded by the merest sip. The behaviour of the males, however, forced me to abandon this idea. They also possess the long beak, and could readily make such perforations if they wished; yet I have never seen one take up his stand upon an acorn and work at it with his augur. Then why this fruitless labour? A mere nothing suffices these abstemious creatures. A superficial operation performed upon the surface of a tender leaf yields them sufficient sustenance.

If the males, the unoccupied males who have leisure to enjoy the pleasures of the palate, ask no more than the sap of the leaf, how should the mothers, busied with the affairs of the breeding-season, find time to waste upon such dearly bought pleasures as the inner juices of the acorn? No, the acorn is not perforated for the purpose of drinking its juices. It is possible that once the beak is deeply sunk, the female may take a mouthful or two, but it is certain that food and drink are not the objects in view.

At last I begin to foresee the solution of the problem. The egg, as I have said, is always at the base of the acorn, in the midst of a soft cottony layer which is moistened by the sap which oozes from the stalk. The grub, upon hatching out, being as yet incapable of attacking the firm substance of the cotyledons, masticates the delicate felt-like layer at the base of the cup and is nourished by its juices.

But as the acorn matures this layer becomes more solid in its consistency. The soft tissues harden; the moist tissues dry up. There is a period during which the acorn fulfils to perfection the conditions most conducive to the welfare of the grub. At an earlier period matters would not have reached the desired stage; at a later period the acorn would be too mature.

The exterior of the acorn gives no indication whatever of the progress of this internal cookery. In order not to inflict unsuitable food on the grub, the mother beetle, not sufficiently informed by the look of the acorn, is thus obliged to taste, at the end of her trunk, the tissues at the base of the cup.

The nurse, before giving her charge a spoonful of broth, tests it by tasting it. In the same way the mother beetle plunges her trunk into the base of the cup, to test the contents before bestowing them upon her offspring. If the food is recognised as being satisfactory the egg is laid; if not, the perforation is abandoned without more ado. This explains the perforations which serve no purpose, in spite of so much labour; the tissues at the base of the cup, being carefully tested, are not found to be in the required condition. The elephant-beetles are difficult to please and take infinite pains when the first mouthful of the grub is in question. To place the egg in a position where the new-born grub will find light and juicy and easily digested nutriment is not enough for those far-seeing mothers; their cares look beyond this point. An intermediary period is desirable, which will lead the little larva from the delicacies of its first hours to the diet of hard acorn. This intermediary period is passed in the gallery, the work of the maternal beak. There it finds the crumbs, the shavings bitten off by the chisels of the rostrum. Moreover, the walls of the tunnel, which are softened by mortification, are better suited than the rest of the acorn to the tender mandibles of the larva.

Before setting to work on the cotyledons the grub does, in fact, commence upon the contents and walls of this tiny passage. It first consumes the shavings lying loose in the passage; it devours the brown fragments adhering to the walls; finally, being now sufficiently strengthened, it attacks the body of the acorn, plunges into it, and disappears. The stomach is ready; the rest is a blissful feast.

This intermediary tunnel must be of a certain length, in order to satisfy the needs of infancy, so the mother must labour at the work of drilling. If the perforation were made solely with the purpose of tasting the material at the base of the acorn and recognising its degree of maturity, the operation might be very much shorter, since the hole could be sunk through the cup itself from a point close to the base. This fact is not unrecognised; I have on occasion found the insect perforating the scaly cup.

In such a proceeding I see the attempt of a gravid mother pressed for time to obtain prompt information. If the acorn is suitable the boring will be recommenced at a more distant point, through the surface of the acorn itself. When an egg is to be laid the rule is to bore the hole from a point as distant as is practicable from the base--as far, in short, as the length of the rostrum will permit.

What is the object of this long perforation, which often occupies more than half the day? Why this tenacious perseverance when, not far from the stalk, at the cost of much less time and fatigue, the rostrum could attain the desired point--the living spring from which the new-born grub is to drink? The mother has her own reasons for toiling in this manner; in doing thus she still attains the necessary point, the base of the acorn, and at the same time--a most valuable result--she prepares for the grub a long tube of fine, easily digested meal.

But these are trivialities! Not so, if you please, but high and important matters, speaking to us of the infinite pains which preside over the preservation of the least of things; witnesses of a superior logic which regulates the smallest details.

The Balaninus, so happily inspired as a mother, has her place in the world and is worthy of notice. So, at least, thinks the blackbird, which gladly makes a meal of the insect with the long beak when fruits grow rare at the end of autumn. It makes a small mouthful, but a tasty, and is a pleasant change after such olives as yet withstand the cold.

And what without the blackbird and its rivalry of song were the reawakening of the woods in spring? Were man to disappear, annihilated by his own foolish errors, the festival of the life-bringing season would be no less worthily observed, celebrated by the fluting of the yellow-billed songster.

To the meritorious role of regaling the blackbird, the minstrel of the forest, the Balaninus adds another--that of moderating the superfluity of vegetation. Like all the mighty who are worthy of their strength, the oak is generous; it produces acorns by the bushel. What could the earth do with such prodigality? The forest would stifle itself for want of room; excess would ruin the necessary.

But no sooner is this abundance of food produced than there is an influx from every side of consumers only too eager to abate this inordinate production. The field-mouse, a native of the woods, stores acorns in a gravel-heap near its hay-lined nest. A stranger, the jay, comes in flocks from far away, warned I know not how. For some weeks it flies feasting from oak to oak, giving vent to its joys and its emotions in a voice like that of a strangling cat; then, its mission accomplished, it returns to the North whence it came.

The Balaninus has anticipated them all. The mother confided her eggs to the acorns while yet they were green. These have now fallen to earth, brown before their time, and pierced by a round hole through which the larva has escaped after devouring the contents. Under one single oak a basket might easily be filled with these ruined shells. More than the jay, more than the field-mouse, the elephant-beetle has contributed to reduce the superfluity of acorns.

Presently man arrives, busied in the interest of his pig. In my village it is quite an important event when the municipal hoardings announce the day for opening the municipal woods for the gathering of acorns. The more zealous visit the woods the day before and select the best places. Next day, at daybreak, the whole family is there. The father beats the upper branches with a pole; the mother, wearing a heavy hempen apron which enables her to force her way through the stubborn undergrowth, gathers those within reach of the hand, while the children collect those scattered upon the ground. First the small baskets are filled, then the big *corbeilles*, and then the sacks.

After the field-mouse, the jay, the weevil, and so many others have taken toll comes man, calculating how many pounds of bacon-fat his harvest will be worth. One regret mingles with the cheer of the occasion; it is to see so many acorns scattered on the ground which are pierced, spoiled, good for nothing. And man curses the author of this destruction; to hear him you would think the forest is meant for him alone, and that the oaks bear acorns only for the sake of his pig.

My friend, I would say to him, the forest guard cannot take legal proceedings against the offender, and it is just as well, for our egoism, which is inclined to see in the acorn only a garland of sausages, would have annoying results. The oak calls the whole world to enjoy its fruits. We take the larger part because we are the stronger. That is our only right.

More important than our rights is the equitable division of the fruits of the earth between the various consumers, great and little, all of whom play their part in this world. If it is good that the blackbird should flute and rejoice in the burgeoning of the spring, then it is no bad thing that acorns should be worm-eaten. In the acorn the dessert of the blackbird is prepared; the Balaninus, the tasty mouthful that puts flesh upon his flanks and music into his throat.

Let the blackbird sing, and let us return to the eggs of the Curculionidae. We know where the egg is--at the base of the acorn, because the tenderest and most juicy tissues of the fruit are there. But how did it get there, so far from the point of entry? A very trifling question, it is true; puerile even, if you will. Do not let us disdain to ask it; science is made of these puerilities.

The first man to rub a piece of amber on his sleeve and to find that it thereupon attracted fragments of chaff had certainly no vision of the electric marvels of our days. He was amusing himself in a childlike manner. Repeated, tested, and probed in every imaginable way, the child's experiment has become one of the forces of the world.

The observer must neglect nothing; for he never knows what may develop out of the humblest fact. So again we will ask: by what process did the egg of the elephant-beetle reach a point so far from the orifice in the acorn?

To one who was not already aware of the position of the egg, but knew that the grub attacked the base of the acorn first, the solution of that fact would be as follows: the egg is laid at the entrance of the tunnel, at the surface, and the grub, crawling down the gallery sunk by the mother, gains of its own accord this distant point where its infant diet is to be found.

Before I had sufficient data this was my own belief; but the mistake was soon exposed. I plucked an acorn just as the mother withdrew, after having for a moment applied the tip of the abdomen to the orifice of the passage just opened by her rostrum. The egg, so it seemed, must be there, at the entrance of the passage.... But no, it was not! It was at the other extremity of the passage! If I dared, I would say it had dropped like a stone into a well.

That idea we must abandon at once; the passage is extremely narrow and encumbered with shavings, so that such a thing would be impossible. Moreover, according to the direction of the stem, accordingly as it pointed upwards or downwards, the egg would have to fall downwards in one acorn and upwards in another.

A second explanation suggests itself, not less perilous. It might be said: "The cuckoo lays her egg on the grass, no matter where; she lifts it in her beak and places it in the nearest appropriate nest." Might not the Balaninus follow an analogous method? Does she employ the rostrum to place the egg in its position at the base of the acorn? I cannot see that the insect has any other implement capable of reaching this remote hiding-place.

Nevertheless, we must hastily reject such an absurd explanation as a last, desperate resort. The elephant-beetle certainly does not lay its egg in the open and seize it in its beak. If it did so the delicate ovum would certainly be destroyed, crushed in the attempt to thrust it down a narrow passage half choked with debris.

This is very perplexing. My embarrassment will be shared by all readers who are acquainted with the structure of the elephant-beetle. The grasshopper has a sabre, an oviscapt which plunges into the earth and sows the eggs at the desired depth; the Leuscopis has a probe which finds its way through the masonry of the mason-bee and lays the egg in the cocoon of the great somnolent larva; but the Balaninus has none of these swords, daggers, or pikes; she has nothing but the tip of her abdomen. Yet she has only to apply that abdominal extremity to the opening of the passage, and the egg is immediately lodged at the very bottom.

Anatomy will give us the answer to the riddle, which is otherwise indecipherable. I open the body of a gravid female. There, before my eyes, is something that takes my breath away. There, occupying the whole length of the body, is an extraordinary device; a red, horny, rigid rod; I had almost said a rostrum, so greatly does it resemble the implement which the insect carries on his head. It is a tube, fine as a horsehair, slightly enlarged at the free extremity, like an old-fashioned blunderbuss, and expanding to form an egg-shaped capsule at the point of origin.

This is the oviduct, and its dimensions are the same as those of the rostrum. As far as the perforating beak can plunge, so far the oviscapt, the interior rostrum, will reach. When working upon her acorn the female chooses the point of attack so that the two complementary instruments can each of them reach the desired point at the base of the acorn.

The matter now explains itself. The work of drilling completed, the gallery ready, the mother turns and places the tip of the abdomen against the orifice. She extrudes the internal mechanism, which easily passes through the loose debris of the boring. No sign of the probe appears, so quickly and discreetly does it work; nor is any trace of it to be seen when, the egg having been properly deposited, the implement ascends and returns to the abdomen. It is over, and the mother departs, and we have not caught a glimpse of her internal mechanism.

Was I not right to insist? An apparently insignificant fact has led to the authentic proof of a fact that the Larinidae had already made me suspect. The long-beaked weevils have an internal probe, an abdominal rostrum, which nothing in their external appearance betrays; they possess, among the hidden organs of the abdomen, the counterpart of the grasshopper's sabre and the ichneumon's dagger.

CHAPTER XVIII
THE PEA-WEEVIL--BRUCHUS PISI

Peas are held in high esteem by mankind. From remote ages man has endeavoured, by careful culture, to produce larger, tenderer, and sweeter varieties. Of an adaptable character, under careful treatment the plant has evolved in a docile fashion, and has ended by giving us what the ambition of the gardener desired. To-day we have gone far beyond the yield of the Varrons and Columelles, and further still beyond the original pea; from the wild seeds confided to the soil by the first man who thought to scratch up the surface of the earth, perhaps with the half-jaw of a cave-bear, whose powerful canine tooth would serve him as a ploughshare!

Where is it, this original pea, in the world of spontaneous vegetation? Our own country has nothing resembling it. Is it to be found elsewhere? On this point botany is silent, or replies only with vague probabilities.

We find the same ignorance elsewhere on the subject of the majority of our alimentary vegetables. Whence comes wheat, the blessed grain which gives us bread? No one knows. You will not find it here, except in the care of man; nor will you find it abroad. In the East, the birthplace of agriculture, no botanist has ever encountered the sacred ear growing of itself on unbroken soil.

Barley, oats, and rye, the turnip and the beet, the beetroot, the carrot, the pumpkin, and so many other vegetable products, leave us in the same perplexity; their point of departure is unknown to us, or at most suspected behind the impenetrable cloud of the centuries. Nature delivered them to us in the full vigour of the thing untamed, when their value as food was indifferent, as to-day she offers us the sloe, the bullace, the blackberry, the crab; she gave them to us in the state of imperfect sketches, for us to fill out and complete; it was for our skill and our labour patiently to induce the nourishing pulp which was the earliest form of capital, whose interest is always increasing in the primordial bank of the tiller of the soil.

As storehouses of food the cereal and the vegetable are, for the greater part, the work of man. The fundamental species, a poor resource in their original state, we borrowed as they were from the natural treasury of the vegetable world; the perfected race, rich in alimentary materials, is the result of our art.

If wheat, peas, and all the rest are indispensable to us, our care, by a just return, is absolutely necessary to them. Such as our needs have made them, incapable of resistance in the bitter struggle for survival, these vegetables, left to themselves without culture, would rapidly disappear, despite the numerical abundance of their seeds, as the foolish sheep would disappear were there no more sheep-folds.

They are our work, but not always our exclusive property. Wherever food is amassed, the consumers collect from the four corners of the sky; they invite themselves to the feast of abundance, and the richer the food the greater their numbers. Man, who alone is capable of inducing agrarian abundance, is by that very fact the giver of an immense banquet at which legions of feasters take their place. By creating more juicy and more generous fruits he calls to his enclosures, despite himself, thousands and thousands of hungry creatures, against whose appetites his prohibitions are helpless. The more he produces, the larger is the tribute demanded of him. Wholesale agriculture and vegetable abundance favour our rival the insect.

This is the immanent law. Nature, with an equal zeal, offers her mighty breast to all her nurslings alike; to those who live by the goods of others no less than to the producers. For us, who plough, sow, and reap, and weary ourselves with labour, she ripens the wheat; she ripens it also for the little Calender-beetle, which, although exempted from the labour of the fields, enters our granaries none the less, and there, with its pointed beak, nibbles our wheat, grain by grain, to the husk.

For us, who dig, weed, and water, bent with fatigue and burned by the sun, she swells the pods of the pea; she swells them also for the weevil, which does no gardener's work, yet takes its share of the harvest at its own hour, when the earth is joyful with the new life of spring.

Let us follow the manoeuvres of this insect which takes its tithe of the green pea. I, a benevolent ratepayer, will allow it to take its dues; it is precisely to benefit it that I have sown a few rows of the beloved plant in a corner of my garden. Without other invitation on my part than this modest expenditure of seed-peas it arrives punctually during the month of May. It has learned that this stony soil, rebellious to the culture of the kitchen-gardener, is bearing peas for the first time. In all haste therefore it has hurried, an agent of the entomological revenue system, to demand its dues.

Whence does it come? It is impossible to say precisely. It has come from some shelter, somewhere, in which it has passed the winter in a state of torpor. The plane-tree, which sheds its rind during the heats of the summer, furnishes an excellent refuge for homeless insects under its partly detached sheets of bark.

I have often found our weevil in such a winter refuge. Sheltered under the dead covering of the plane, or otherwise protected while the winter lasts, it awakens from its torpor at the first touch of a kindly sun. The almanack of the instincts has aroused it; it knows as well as the gardener when the pea-vines are in flower, and seeks its favourite plant, journeying thither from every side, running with quick, short steps, or nimbly flying.

A small head, a fine snout, a costume of ashen grey sprinkled with brown, flattened wing-covers, a dumpy, compact body, with two large black dots on the rear segment--such is the summary portrait of my visitor. The middle of May approaches, and with it the van of the invasion.

They settle on the flowers, which are not unlike white-winged butterflies. I see them at the base of the blossom or inside the cavity of the "keel" of the flower, but the majority explore the petals and take possession of them. The time for laying the eggs has not yet arrived. The morning is mild; the sun is warm without being oppressive. It is the moment of nuptial flights; the time of rejoicing in the splendour of the sunshine. Everywhere are creatures rejoicing to be alive. Couples come together, part, and re-form. When towards noon the heat becomes too great, the weevils retire into the shadow, taking refuge singly in the folds of the flowers whose secret corners they know so well. To-morrow will be another day of festival, and the next day also, until the pods, emerging from the shelter of the "keel" of the flower, are plainly visible, enlarging from day to day.

A few gravid females, more pressed for time than the others, confide their eggs to the growing pod, flat and meagre as it issues from its floral sheath. These hastily laid batches of eggs, expelled perhaps by the exigencies of an ovary incapable of further delay, seem to me in serious danger; for the seed in which the grub must establish itself is as yet no more than a tender speck of green, without firmness and without any farinaceous tissue. No larva could possible find sufficient nourishment there, unless it waited for the pea to mature.

But is the grub capable of fasting for any length of time when once hatched? It is doubtful. The little I have seen tells me that the new-born grub must establish itself in the midst of its food as quickly as possible, and that it perishes unless it can do so. I am therefore of opinion that such eggs as are deposited in immature pods are lost. However, the race will hardly suffer by such a loss, so fertile is the little beetle. We shall see directly how prodigal the female is of her eggs, the majority of which are destined to perish.

The important part of the maternal task is completed by the end of May, when the shells are swollen by the expanding peas, which have reached their final growth, or are but little short of it. I was anxious to see the female Bruchus at work in her quality of Curculionid, as our classification declares her.[8] The other weevils are Rhyncophora, beaked insects, armed with a drill with which to prepare the hole in which the egg is laid. The Bruchus possesses only a short snout or muzzle, excellently adapted for eating soft tissues, but valueless as a drill.

The method of installing the family is consequently absolutely different. There are no industrious preparations as with the Balinidae, the Larinidae, and the Rhynchitides. Not being equipped with a long oviscapt, the mother sows her eggs in the open, with no protection against the heat of the sun and the variations of temperature. Nothing could be simpler, and nothing more perilous to the eggs, in the absence of special characteristics which would enable them to resist the alternate trials of heat and cold, moisture and drought.

In the caressing sunlight of ten o'clock in the morning the mother runs up and down the chosen pod, first on one side, then on the other, with a jerky, capricious, unmethodical gait. She repeatedly extrudes a short oviduct, which oscillates right and left as though to graze the skin of the pod. An egg follows, which is abandoned as soon as laid.

A hasty touch of the oviduct, first here, then there, on the green skin of the pea-pod, and that is all. The egg is left there, unprotected, in the full sunlight. No choice of position is made such as might assist the grub when it seeks to penetrate its larder. Some eggs are laid on the swellings created by the peas beneath; others in the barren valleys which separate them. The first are close to the peas, the second at some distance from them. In short, the eggs of the Bruchus are laid at random, as though on the wing.

We observe a still more serious vice: the number of eggs is out of all proportion to the number of peas in the pod. Let us note at the outset that each grub requires one pea; it is the necessary ration, and is largely sufficient to one larva, but is not enough for several, nor even for two. One pea to each grub, neither more nor less, is the unchangeable rule.

We should expect to find signs of a procreative economy which would impel the female to take into account the number of peas contained in the pod which she has just explored; we might expect her to set a numerical limit on her eggs in conformity with that of the peas available. But no such limit is observed. The rule of one pea to one grub is always contradicted by the multiplicity of consumers.

My observations are unanimous on this point. The number of eggs deposited on one pod always exceeds the number of peas available, and often to a scandalous degree. However meagre the contents of the pod there is a superabundance of consumers. Dividing the sum of the eggs upon such or such a pod by that of the peas contained therein, I find there are five to eight claimants for each pea; I have found ten, and there is no reason why this prodigality should not go still further. Many are called, but few are chosen! What is to become of all these supernumeraries, perforce excluded from the banquet for want of space?

The eggs are of a fairly bright amber yellow, cylindrical in form, smooth, and rounded at the ends. Their length is at most a twenty-fifth of an inch. Each is affixed to the pod by means of a slight network of threads of coagulated albumen. Neither wind nor rain can loosen their hold.

The mother not infrequently emits them two at a time, one above the other; not infrequently, also, the uppermost of the two eggs hatches before the other, while the latter fades and perishes. What was lacking to this egg, that it should fail to produce a grub? Perhaps a bath of sunlight; the incubating heat of which the outer egg has robbed it. Whether on account of the fact that it is shadowed by the other egg, or for other reasons, the elder of the eggs in a group of two rarely follows the normal course, but perishes on the pod, dead without having lived.

There are exceptions to this premature end; sometimes the two eggs develop equally well; but such cases are exceptional, so that the Bruchid family would be reduced to about half its dimensions if the binary system were the rule. To the detriment of our peas and to the advantage of the beetle, the eggs are commonly laid one by one and in isolation.

A recent emergence is shown by a little sinuous ribbon-like mark, pale or whitish, where the skin of the pod is raised and withered, which starts from the egg and is the work of the new-born larva; a sub-epidermic tunnel along which the grub works its way, while seeking a point from which it can escape into a pea. This point once attained, the larva, which is scarcely a twenty-fifth of an inch in length, and is white with a black head, perforates the envelope and plunges into the capacious hollow of the pod.

It has reached the peas and crawls upon the nearest. I have observed it with the magnifier. Having explored the green globe, its new world, it begins to sink a well perpendicularly into the sphere. I have often seen it half-way in, wriggling its tail in the effort to work the quicker. In a short time the grub disappears and is at home. The point of entry, minute, but always easily recognisable by its brown coloration on the pale green background of the pea, has no fixed location; it may be at almost any point on the surface of the pea, but an exception is usually made of the lower half; that is, the hemisphere whose pole is formed by the supporting stem.

It is precisely in this portion that the germ is found, which will not be eaten by the larva, and will remain capable of developing into a plant, in spite of the large aperture made by the emergence of the adult insect. Why is this particular portion left untouched? What are the motives that safeguard the germ?

It goes without saying that the Bruchus is not considering the gardener. The pea is meant for it and for no one else. In refusing the few bites that would lead to the death of the seed, it has no intention of limiting its destruction. It abstains from other motives.

Let us remark that the peas touch laterally, and are pressed one against the other, so that the grub, when searching for a point of attack, cannot circulate at will. Let us also note that the lower pole expands into the umbilical excrescence, which is less easy of perforation than those parts protected by the skin alone. It is even possible that the umbilicum, whose organisation differs from that of the rest of the pea, contains a peculiar sap that is distasteful to the little grub.

Such, doubtless, is the reason why the peas exploited by the Bruchus are still able to germinate. They are damaged, but not dead, because the invasion was conducted from the free hemisphere, a portion less vulnerable and more easy of access. Moreover, as the pea in its entirety is too large for a single grub to consume, the consumption is limited to the portion preferred by the consumer, and this portion is not the essential portion of the pea.

With other conditions, with very much smaller or very much larger seeds, we shall observe very different results. If too small, the germ will perish, gnawed like the rest by the insufficiently provisioned inmate; if too large, the abundance of food will permit of several inmates. Exploited in the absence of the pea, the cultivated vetch and the broad bean afford us an excellent example; the smaller seed, of which all but the skin is devoured, is left incapable of germination; but the large bean, even though it may have held a number of grubs, is still capable of sprouting.

Knowing that the pod always exhibits a number of eggs greatly in excess of the enclosed peas, and that each pea is the exclusive property of one grub, we naturally ask what becomes of the superfluous grubs. Do they perish outside when the more precocious have one by one taken their places in their vegetable larder? or do they succumb to the intolerant teeth of the first occupants? Neither explanation is correct. Let us relate the facts.

On all old peas--they are at this stage dry--from which the adult Bruchus has emerged, leaving a large round hole of exit, the magnifying-glass will show a variable number of fine reddish punctuations, perforated in the centre. What are these spots, of which I count five, six, and even more on a single pea? It is impossible to be mistaken: they are the points of entry of as many grubs. Several grubs have entered the pea, but of the whole group only one has survived, fattened, and attained the adult age. And the others? We shall see.

At the end of May, and in June, the period of egg-laying, let us inspect the still green and tender peas. Nearly all the peas invaded show us the multiple perforations already observed on the dry peas abandoned by the weevils. Does this actually mean that there are several grubs in the pea? Yes. Skin the peas in question, separate the cotyledons, and break them up as may be necessary. We shall discover several grubs, extremely youthful, curled up comma-wise, fat and lively, each in a little round niche in the body of the pea.

Peace and welfare seem to reign in the little community. There is no quarrelling, no jealousy between neighbours. The feast has commenced; food is abundant, and the feasters are separated one from another by the walls of uneaten substance. With this isolation in separate cells no conflicts need be feared; no sudden bite of the mandibles, whether intentional or accidental. All the occupants enjoy the same rights of property, the same appetite, and the same strength. How does this communal feast terminate?

Having first opened them, I place a number of peas which are found to be well peopled in a glass test-tube. I open others daily. In this way I keep myself informed as to the progress of the various larvae. At first nothing noteworthy is to be seen. Isolated in its narrow chamber, each grub nibbles the substance around it, peacefully and parsimoniously. It is still very small; a mere speck of food is a feast; but the contents of one pea will not suffice the whole number to the end. Famine is ahead, and all but one must perish.

Soon, indeed, the aspect of things is entirely changed. One of the grubs--that which occupies the central position in the pea--begins to grow more quickly than the others. Scarcely has it surpassed the others in size when the latter cease to eat, and no longer attempt to burrow forwards. They lie motionless and resigned; they die that gentle death which comes to unconscious lives. Henceforth the entire pea belongs to the sole survivor. Now what has happened that these lives around the privileged one should be thus annihilated? In default of a satisfactory reply, I will propose a suggestion.

In the centre of the pea, less ripened than the rest of the seed by the chemistry of the sun, may there not be a softer pulp, of a quality better adapted to the infantile digestion of the grub? There, perhaps, being nourished by tenderer, sweeter, and perhaps more tasty tissues, the stomach becomes more vigorous, until it is fit to undertake less easily digested food. A nursling is fed on milk before proceeding to bread and broth. May not the central portion of the pea be the feeding-bottle of the Bruchid?

With equal rights, fired by an equal ambition, all the occupants of the pea bore their way towards the delicious morsel. The journey is laborious, and the grubs must rest frequently in their provisional niches. They rest; while resting they frugally gnaw the riper tissues surrounding them; they gnaw rather to open a way than to fill their stomachs.

Finally one of the excavators, favoured by the direction taken, attains the central portion. It establishes itself there, and all is over; the others have only to die. How are they warned that the place is taken? Do they hear their brother gnawing at the walls of his lodging? can they feel the vibration set up by his nibbling mandibles? Something of the kind must happen, for from that moment they make no attempt to burrow further. Without struggling against the fortunate winner, without seeking to dislodge him, those which are beaten in the race give themselves up to death. I admire this candid resignation on the part of the departed.

Another condition--that of space--is also present as a factor. The pea-weevil is the largest of our Bruchidae. When it attains the adult stage it requires a certain amplitude of lodging, which the other weevils do not require in the same degree. A pea provides it with a sufficiently spacious cell; nevertheless, the cohabitation of two in one pea would be impossible; there would be no room, even were the two to put up with a certain discomfort. Hence the necessity of an inevitable decimation, which will suppress all the competitors save one.

Now the superior volume of the broad bean, which is almost as much beloved by the weevil as the pea, can lodge a considerable community, and the solitary can live as a cenobite. Without encroaching on the domain of their neighbours, five or six or more can find room in the one bean.

Moreover, each grub can find its infant diet; that is, that layer which, remote from the surface, hardens only gradually and remains full of sap until a comparatively late period. This inner layer represents the crumb of a loaf, the rest of the bean being the crust.

In the pea, a sphere of much less capacity, it occupies the central portion; a limited point at which the grub develops, and lacking which it perishes; but in the bean it lines the wide adjoining faces of the two flattened cotyledons. No matter where the point of attack is made, the grub has only to bore straight down when it quickly reaches the softer tissues. What is the result? I have counted the eggs adhering to a bean-pod and the beans included in the pod, and comparing the two figures I find that there is plenty of room for the whole family at the rate of five or six dwellers in each bean. No superfluous larvae perish of hunger when barely issued from the egg; all have their share of the ample provision; all live and prosper. The abundance of food balances the prodigal fertility of the mother.

If the Bruchus were always to adopt the broad bean for the establishment of her family I could well understand the exuberant allowance of eggs to one pod; a rich food-stuff easily obtained evokes a large batch of eggs. But the case of the pea perplexes me. By what aberration does the mother abandon her children to starvation on this totally insufficient vegetable? Why so many grubs to each pea when one pea is sufficient only for one grub?

Matters are not so arranged in the general balance-sheet of life. A certain foresight seems to rule over the ovary so that the number of mouths is in proportion to the abundance or scarcity of the food consumed. The Scarabaeus, the Sphex, the Necrophorus, and other insects which prepare and preserve alimentary provision for their families, are all of a narrowly limited fertility, because the balls of dung, the dead or paralysed insects, or the buried corpses of animals on which their offspring are nourished are provided only at the cost of laborious efforts.

The ordinary bluebottle, on the contrary, which lays her eggs upon butcher's meat or carrion, lays them in enormous batches. Trusting in the inexhaustible riches represented by the corpse, she is prodigal of offspring, and takes no account of numbers. In other cases the provision is acquired by audacious brigandage, which exposes the newly born offspring to a thousand mortal accidents. In such cases the mother balances the chances of destruction by an exaggerated flux of eggs. Such is the case with the Meloides, which, stealing the goods of others under conditions of the greatest peril, are accordingly endowed with a prodigious fertility.

The Bruchus knows neither the fatigues of the laborious, obliged to limit the size of her family, nor the misfortunes of the parasite, obliged to produce an exaggerated number of offspring. Without painful search, entirely at her ease, merely moving in the sunshine over her favourite plant, she can ensure a sufficient provision for each of her offspring; she can do so, yet is foolish enough to over-populate the pod of the pea; a nursery insufficiently provided, in which the great majority will perish of starvation. This ineptitude is a thing I cannot understand: it clashes too completely with the habitual foresight of the maternal instinct.

I am inclined to believe that the pea is not the original food plant of the Bruchus. The original plant must rather have been the bean, one seed of which is capable of supporting half a dozen or more larvae. With the larger cotyledon the crying disproportion between the number of eggs and the available provision disappears.

Moreover, it is indubitable that the bean is of earlier date than the pea. Its exceptional size and its agreeable flavour would certainly have attracted the attention of man from the remotest periods. The bean is a ready-made mouthful, and would be of the greatest value to the hungry tribe. Primitive man would at an early date have sown it beside his wattled hut. Coming from Central Asia by long stages, their wagons drawn by shaggy oxen and rolling on the circular discs cut from the trunks of trees, the early immigrants would have brought to our virgin land, first the bean, then the pea, and finally the cereal, that best of safeguards against famine. They taught us the care of herds, and the use of bronze, the material of the first metal implement. Thus the dawn of civilisation arose over France. With the bean did those ancient teachers also involuntarily bring us the insect which to-day disputes it with us? It is doubtful; the Bruchidae seem to be indigenous. At all events, I find them levying tribute from various indigenous plants, wild vegetables which have never tempted the appetite of man. They abound in particular upon the great forest vetch (Lathyrus latifolius), with its magnificent heads of flowers and long handsome pods. The seeds are not large, being indeed smaller than the garden pea; but eaten to the very skin, as they invariably are, each is sufficient to the needs of its grub.

We must not fail to note their number. I have counted more than twenty in a single pod, a number unknown in the case of the pea, even in the most prolific varieties. Consequently this superb vetch is in general able to nourish without much loss the family confided to its pod.

Where the forest vetch is lacking, the Bruchus, none the less, bestows its habitual prodigality of eggs upon another vegetable of similar flavour, but incapable of nourishing all the grubs: for the example, the travelling vetch (Vicia peregrina) or the cultivated vetch (Vicia sativa). The number of eggs remains high even upon insufficient pods, because the original food-plant offered a copious provision, both in the multiplicity and the size of the seeds. If the Bruchus is really a stranger, let us regard the bean as the original food-plant; if indigenous, the large vetch.

Sometime in the remote past we received the pea, growing it at first in the prehistoric vegetable garden which already supplied the bean. It was found a better article of diet than the broad bean, which to-day, after such good service, is comparatively neglected. The weevil was of the same opinion as man, and without entirely forgetting the bean and the vetch it established the greater part of its tribe upon the pea, which from century to century was more widely cultivated. To-day we have to share our peas: the Bruchidae take what they need, and bestow their leavings on us.

This prosperity of the insect which is the offspring of the abundance and quality of our garden products is from another point of view equivalent to decadence. For the weevil, as for ourselves, progress in matters of food and drink is not always beneficial. The race would profit better if it remained frugal. On the bean and the vetch the Bruchus founded colonies in which the infant mortality was low. There was room for all. On the pea-vine, delicious though its fruits may be, the greater part of its offspring die of starvation. The rations are few, and the hungry mouths are multitudinous.

We will linger over this problem no longer. Let us observe the grub which has now become the sole tenant of the pea by the death of its brothers. It has had no part in their death; chance has favoured it, that is all. In the centre of the pea, a wealthy solitude, it performs the duty of a grub; the sole duty of eating. It nibbles the walls enclosing it, enlarging its lodgment, which is always entirely filled by its corpulent body. It is well shaped, fat, and shining with health. If I disturb it, it turns gently in its niche and sways its head. This is its manner of complaining of my importunities. Let us leave it in peace.

It profits so greatly and so swiftly by its position that by the time the dog-days have come it is already preparing for its approaching liberation. The adult is not sufficiently well equipped to open for itself a way out through the pea, which is now completely hardened. The larva knows of this future helplessness, and with consummate art provides for its release. With its powerful mandibles it bores a channel of exit, exactly round, with extremely clean-cut sides. The most skilful ivory-carver could do no better.

To prepare the door of exit in advance is not enough; the grub must also provide for the tranquillity essential to the delicate processes of nymphosis. An intruder might enter by the open door and injure the helpless nymph. This passage must therefore remain closed. But how?

As the grub bores the passage of exit it consumes the farinaceous matter without leaving a crumb. Having come to the skin of the pea it stops short. This membrane, semi-translucid, is the door to the chamber of metamorphosis, its protection against the evil intentions of external creatures.

It is also the only obstacle which the adult will encounter at the moment of exit. To lessen the difficulty of opening it the grub takes the precaution of gnawing at the inner side of the skin, all round the circumference, so as to make a line of least resistance. The perfect insect will only have to heave with its shoulder and strike a few blows with its head in order to raise the circular door and knock it off like the lid of a box. The passage of exit shows through the diaphanous skin of the pea as a large circular spot, which is darkened by the obscurity of the interior. What passes behind it is invisible, hidden as it is behind a sort of ground glass window.

A pretty invention, this little closed porthole, this barricade against the invader, this trap-door raised by a push when the time has come for the hermit to enter the world. Shall we credit it to the Bruchus? Did the ingenious insect conceive the undertaking? Did it think out a plan and work out a scheme of its own devising? This would be no small triumph for the brain of a weevil. Before coming to a conclusion let us try an experiment.

I deprive certain occupied peas of their skin, and I dry them with abnormal rapidity, placing them in glass test tubes. The grubs prosper as well as in the intact peas. At the proper time the preparations for emergence are made.

If the grub acts on its own inspiration, if it ceases to prolong its boring directly it recognises that the outer coating, auscultated from time to time, is sufficiently thin, what will it do under the conditions of the present test? Feeling itself at the requisite distance from the surface it will stop boring; it will respect the outer layer of the bare pea, and will thus obtain the indispensable protecting screen.

Nothing of the kind occurs. In every case the passage is completely excavated; the entrance gapes wide open, as large and as carefully executed as though the skin of the pea were in its place. Reasons of security have failed to modify the usual method of work. This open lodging has no defence against the enemy; but the grub exhibits no anxiety on this score.

Neither is it thinking of the outer enemy when it bores down to the skin when the pea is intact, and then stops short. It suddenly stops because the innutritious skin is not to its taste. We ourselves remove the parchment-like skins from a mess of pease-pudding, as from a culinary point of view they are so much waste matter. The larva of the Bruchus, like ourselves, dislikes the skin of the pea. It stops short at the horny covering, simply because it is checked by an uneatable substance. From this aversion a little miracle arises; but the insect has no sense of logic; it is passively obedient to the superior logic of facts. It obeys its instinct, as unconscious of its act as is a crystal when it assembles, in exquisite order, its battalions of atoms.

Sooner or later during the month of August we see a shadowy circle form on each inhabited pea; but only one on each seed. These circles of shadow mark the doors of exit. Most of them open in September. The lid, as though cut out with a punch, detaches itself cleanly and falls to the ground, leaving the orifice free. The Bruchus emerges, freshly clad, in its final form.

The weather is delightful. Flowers are abundant, awakened by the summer showers; and the weevils visit them in the lovely autumn weather. Then, when the cold sets in, they take up their winter quarters in any suitable retreat. Others, still numerous, are less hasty in quitting the native seed. They remain within during the whole winter, sheltered behind the trap-door, which they take care not to touch. The door of the cell will not open on its hinges, or, to be exact, will not yield along the line of least resistance, until the warm days return. Then the late arrivals will leave their shelter and rejoin the more impatient, and both will be ready for work when the pea-vines are in flower.

To take a general view of the instincts in their inexhaustible variety is, for the observer, the great attraction of the entomological world; for nowhere do we gain a clearer sight of the wonderful way in which the processes of life are ordered. Thus regarded entomology is not, I know, to the taste of everybody; the simple creature absorbed in the doings and habits of insects is held in low esteem. To the terrible utilitarian, a bushel of peas preserved from the weevil is of more importance than a volume of observations which bring no immediate profit.

Yet who has told you, O man of little faith, that what is useless to-day will not be useful to-morrow? If we learn the customs of insects or animals we shall understand better how to protect our goods. Do not despise disinterested knowledge, or you may rue the day. It is by the accumulation of ideas, whether immediately applicable or otherwise, that humanity has done, and will continue to do, better to-day than yesterday, and better to-morrow than to-day. If we live on peas and beans, which we dispute with the weevil, we also live by knowledge, that mighty kneading-trough in which the bread of progress is mixed and leavened. Knowledge is well worth a few beans.

Among other things, knowledge tells us: "The seedsman need not go to the expense of waging war upon the weevil. When the peas arrive in the granary, the harm is already done; it is irreparable, but not transmissible. The untouched peas have nothing to fear from the neighbourhood of those which have been attacked, however long the mixture is left. From the latter the weevils will issue when their time has come; they will fly away from the storehouse if escape is possible; if not, they will perish without in any way attacking the sound peas. No eggs, no new generation will ever be seen upon or within the dried peas in the storehouse; there the adult weevil can work no further mischief."

The Bruchus is not a sedentary inhabitant of granaries: it requires the open air, the sun, the liberty of the fields. Frugal in everything, it absolutely disdains the hard tissues of the vegetable; its tiny mouth is content with a few honeyed mouthfuls, enjoyed upon the flowers. The larvae, on the other hand, require the tender tissues of the green pea growing in the pod. For these reasons the granary knows no final multiplication on the part of the despoiler.

The origin of the evil is in the kitchen-garden. It is there that we ought to keep a watch on the misdeeds of the Bruchus, were it not for the fact that we are nearly always weaponless when it comes to fighting an insect. Indestructible by reason of its numbers, its small size, and its cunning, the little creature laughs at the anger of man. The gardener curses it, but the weevil is not disturbed: it imperturbably continues its trade of levying tribute. Happily we have assistants more patient and more clear-sighted than ourselves.

During the first week of August, when the mature Bruchus begins to emerge, I notice a little Chalcidian, the protector of our peas. In my rearing-cages it issues under my eyes in abundance from the peas infested by the grub of the weevil. The female has a reddish head and thorax; the abdomen is black, with a long augur-like oviscapt. The male, a little smaller, is black. Both sexes have reddish claws and thread-like antennae.

In order to escape from the pea the slayer of the weevil makes an opening in the centre of the circular trap-door which the grub of the weevil prepared in view of its future deliverance. The slain has prepared the way for the slayer. After this detail the rest may be divined.

When the preliminaries to the metamorphosis are completed, when the passage of escape is bored and furnished with its lid of superficial membrane, the female Chalcidian arrives in a busy mood. She inspects the peas, still on the vine, and enclosed in their pods; she auscultates them with her antennae; she discovers, hidden under the general envelope, the weak points in the epidermic covering of the peas. Then, applying her oviscapt, she thrusts it through the side of the pod and perforates the circular trap-door. However far withdrawn into the centre of the pea, the Bruchus, whether larvae or nymph, is reached by the long oviduct. It receives an egg in its tender flesh, and the thing is done. Without possibility of defence, since it is by now a somnolent grub or a helpless pupa, the embryo weevil is eaten until nothing but skin remains. What a pity that we cannot at will assist the multiplication of this eager exterminator! Alas! our assistants have got us in a vicious circle, for if we wished to obtain the help of any great number of Chalcidians we should be obliged in the first place to breed a multiplicity of Bruchidae.

CHAPTER XIX
AN INVADER.--THE HARICOT-WEEVIL

If there is one vegetable on earth that more than any other is a gift of the gods, it is the haricot bean. It has all the virtues: it forms a soft paste upon the tongue; it is extremely palatable, abundant, inexpensive, and highly nutritious. It is a vegetable meat which, without being bloody and repulsive, is the equivalent of the horrors outspread upon the butcher's slab. To recall its services the more emphatically, the Provencal idiom calls it the ***gounflo-gus***--the filler of the poor.

Blessed Bean, consoler of the wretched, right well indeed do you fill the labourer, the honest, skilful worker who has drawn a low number in the crazy lottery of life. Kindly Haricot, with three drops of oil and a dash of vinegar you were the favourite dish of my young years; and even now, in the evening of my days, you are welcome to my humble porringer. We shall be friends to the last.

To-day it is not my intention to sing your merits; I wish simply to ask you a question, being curious: What is the country of your origin? Did you come from Central Asia with the broad bean and the pea? Did you make part of that collection of seeds which the first pioneers of culture brought us from their gardens? Were you known to antiquity?

Here the insect, an impartial and well-informed witness, answers: "No; in our country antiquity was not acquainted with the haricot. The precious vegetable came hither by the same road as the broad bean. It is a foreigner, and of comparatively recent introduction into Europe."

The reply of the insect merits serious examination, supported as it is by extremely plausible arguments. Here are the facts. For years attentive to matters agricultural, I had never seen haricots attacked by any insect whatever; not even by the Bruchidae, the licensed robbers of leguminous seeds.

On this point I have questioned my peasant neighbours. They are men of the extremest vigilance in all that concerns their crops. To steal their property is an abominable crime, swiftly discovered. Moreover, the housewife, who individually examines all beans intended for the saucepan, would inevitably find the malefactor.

All those I have spoken to replied to my questions with a smile in which I read their lack of faith in my knowledge of insects. "Sir," they said, "you must know that there are never grubs in the haricot bean. It is a blessed vegetable, respected by the weevil. The pea, the broad bean, the vetch, and the chick-pea all have their vermin; but the haricot, *lou gounflo-gus*, never. What should we do, poor folk as we are, if the *Courcoussoun* robbed us of it?"

The fact is that the weevil despises the haricot; a very curious dislike if we consider how industriously the other vegetables of the same family are attacked. All, even the beggarly lentil, are eagerly exploited; whilst the haricot, so tempting both as to size and flavour, remains untouched. It is incomprehensible. Why should the Bruchus, which without hesitation passes from the excellent to the indifferent, and from the indifferent to the excellent, disdain this particularly toothsome seed? It leaves the forest vetch for the pea, and the pea for the broad bean, as pleased with the small as with the large, yet the temptations of the haricot bean leave it indifferent. Why?

Apparently because the haricot is unknown to it. The other leguminous plants, whether native or of Oriental origin, have been familiar to it for centuries; it has tested their virtues year by year, and, confiding in the lessons of the past, it bases its forethought for the future upon ancient custom. The haricot is avoided as a newcomer, whose merits it has not yet learned.

The insect emphatically informs us that with us the haricot is of recent date. It has come to us from a distant country: and assuredly from the New World. Every edible vegetable attracts its consumers. If it had originated in the Old World the haricot would have had its licensed consumers, as have the pea, the lentil, and the broad bean. The smallest leguminous seed, if barely bigger than a pin's head, nourishes its weevil; a dwarf which patiently nibbles it and excavates a dwelling; but the plump, delicious haricot is spared.

This astonishing immunity can have only one explanation: like the potato and the maize-plant, the haricot is a gift of the New World. It arrived in Europe without the company of the insect which exploits it in its native country; it has found in our fields another world of insects, which have despised it because they did not know it. Similarly the potato and the ear of maize are untouched in France unless their American consumers are accidentally imported with them.

The verdict of the insect is confirmed by the negative testimony of the ancient classics; the haricot never appears on the table of the Greek or Roman peasant. In the second Eclogue of Virgil Thestylis prepares the repast of the harvesters:--

Thestylis et rapido fessis messoribus aestu   Allia serpyllumque herbas contundit olentes.

This mixture is the equivalent of the *aioli*, dear to the Provencal palate. It sounds very well in verse, but is not very substantial. On such an occasion men would look for that fundamental dish, the plate of red haricots, seasoned with chopped onions. All in good time; this at least would ballast the stomach. Thus refreshed in the open air, listening to the song of the cigales, the gang of harvesters would take their mid-day rest and gently digest their meal in the shadows of the sheaves. Our modern Thestylis, differing little from her classic sister, would take good care not to forget the *gounflo-gus*, that economical resource of large appetites. The Thestylis of the past did not think of providing it because she did not know it.

The same author shows us Tityrus offering a night's hospitality to his friend Meliboeus, who has been driven from his property by the soldiers of Octavius, and goes limping behind his flock of goats. We shall have, says Tityrus, chestnuts, cheese, and fruits. History does not say if Meliboeus allowed himself to be tempted. It is a pity; for during the frugal meal we might have learned in a more explicit fashion that the shepherds of the ancient world were not acquainted with the haricot.

Ovid tells us, in a delightful passage, of the manner in which Philemon and Baucis received the gods unawares as guests in their humble cottage. On the three-legged table, which was levelled by means of a potsherd under one of the legs, they served cabbage soup, rusty bacon, eggs poached for a minute in the hot cinders, cornel-berries pickled in brine, honey, and fruits. In this rustic abundance one dish was lacking; an essential dish, which the Baucis of our countryside would never forget. After bacon soup would follow the obligatory plate of haricots. Why did Ovid, so prodigal of detail, neglect to mention a dish so appropriate to the occasion? The reply is the same as before: because he did not know of it.

In vain have I recapitulated all that my reading has taught me concerning the rustic dietary of ancient times; I can recollect no mention of the haricot. The worker in the vineyard and the harvester have their lupins, broad beans, peas, and lentils, but never the bean of beans, the haricot.

The haricot has a reputation of another kind. It is a source of flatulence; you eat it, as the saying is, and then you take a walk. It lends itself to the gross pleasantries loved of the populace; especially when they are formulated by the shameless genius of an Aristophanes or a Plautus. What merriment over a simple allusion to the sonorous bean, what guffaws from the throats of Athenian sailors or Roman porters! Did the two masters, in the unfettered gaiety of a language less reserved than our own, ever mention the virtues of the haricot? No; they are absolutely silent concerning the trumpet-voiced vegetable.

The name of the bean is a matter for reflection. It is of an unfamiliar sound, having no affinity with our language. By its unlikeness to our native combinations of sounds, it makes one think of the West Indies or South America, as do **caoutchouc** and **cacao**. Does the word as a matter of fact come from the American Indians? Did we receive, together with the vegetable, the name by which it is known in its native country? Perhaps; but how are we to know? Haricot, fantastic haricot, you set us a curious philological problem.

It is also known in French as **faseole**, or **flageolet**. The Provencal calls it **faiou** and **faviou**; the Catalan, **fayol**; the Spaniard, **faseolo**; the Portuguese, **feyao**; the Italian, **fagiuolo**. Here I am on familiar ground: the languages of the Latin family have preserved, with the inevitable modifications, the ancient word **faseolus**.

Now, if I consult my dictionary I find: **faselus, faseolus, phaseolus**, haricot. Learned lexicographer, permit me to remark that your translation is incorrect: **faselus, faseolus** cannot mean haricot. The incontestable proof is in the Georgics, where Virgil tells us at what season we must sow the **faselus**. He says:--

     Si vero viciamque seres vilemque faselum ...   Haud obscura cadens mittet tibi signa Bootes;   Incipe, et ad medias sementem extende pruinas.

Nothing is clearer than the precept of the poet who was so admirably familiar with all matters agricultural; the sowing of the **faselus** must be commenced when the constellation of Bootes disappears at the set of sun, that is, in October; and it is to be continued until the middle of the winter.

These conditions put the haricot out of the running: it is a delicate plant, which would never survive the lightest frost. Winter would be fatal to it, even under Italian skies. More refractory to cold on account of the country of their origin, peas, broad beans, and vetches, and other leguminous plants have nothing to fear from an autumn sowing, and prosper during the winter provided the climate be fairly mild.

What then is represented by the *faselus* of the Georgics, that problematical vegetable which has transmitted its name to the haricot in the Latin tongues? Remembering that the contemptuous epithet *vilis* is used by the poet in qualification, I am strongly inclined to regard it as the cultivated vetch, the big square pea, the little-valued *jaisso* of the Provencal peasant.

The problem of the haricot stood thus, almost elucidated by the testimony of the insect world alone, when an unexpected witness gave me the last word of the enigma. It was once again a poet, and a famous poet, M. Jose-Maria de Heredia, who came to the aid of the naturalist. Without suspecting the service he was rendering, a friend of mine, the village schoolmaster, lent me a magazine[9] in which I read the following conversation between the master-sonneteer and a lady journalist, who was anxious to know which of his own works he preferred.

"What would you have me say?" said the poet.

"I do not know what to say, I do not know which sonnet I prefer; I have taken horrible pains with all of them.... But you, which do you prefer?"

"My dear master, how can I choose out of so many jewels, when each one is perfect in its beauty? You flash pearls, emeralds, and rubies before my astonished eyes: how should I decide to prefer the emerald to the pearl? I am transported by admiration of the whole necklace."

"Well, as for me, there is something I am more proud of than of all my sonnets, and which has done much more for my reputation than my verses."

I opened my eyes wide, "What is that?" I asked. The master looked at me mischievously; then, with that beautiful light in his eyes which fires his youthful countenance, he said triumphantly--

"It is my discovery of the etymology of the word haricot!"

I was so amazed that I forgot to laugh.

"I am perfectly serious in telling you this."

"I know, my dear master, of your reputation for profound scholarship: but to imagine, on that account, that you were famed for your discovery of the etymology of haricot--I should never have expected it! Will you tell me how you made the discovery?"

"Willingly. See now: I found some information respecting the haricot while studying that fine seventeenth-century work of natural history by Hernandez: *De Historia plantarum novi orbis*. The word haricot was unknown in France until the seventeenth century: people used the word *feve* or *phaseol*: in Mexican, *ayacot*. Thirty species of haricot were cultivated in Mexico before the conquest. They are still known as *ayacot*, especially the red haricot, spotted with black or violet. One day at the house of Gaston Paris I met a famous scholar. Hearing my name, he rushed at me and asked if it was I who had discovered the etymology of the word haricot. He was absolutely ignorant of the fact that I had written verses and published the *Trophees*."--

A very pretty whim, to count the jewellery of his famous sonnets as second in importance to the nomenclature of a vegetable! I in my turn was delighted with his *ayacot*. How right I was to suspect the outlandish word of American Indian origin! How right the insect was, in testifying, in its own fashion, that the precious bean came to us from the New World! While still retaining its original name--or something sufficiently like it--the bean of Montezuma, the Aztec *ayacot*, has migrated from Mexico to the kitchen-gardens of Europe.

But it has reached us without the company of its licensed consumer; for there must assuredly be a weevil in its native country which levies tribute on its nourishing tissues. Our native bean-eaters have mistaken the stranger; they have not had time as yet to grow familiar with it, or to appreciate its merits; they have prudently abstained from touching the *ayacot*, whose novelty awoke suspicion. Until our own days the Mexican bean remained untouched: unlike our other leguminous seeds, which are all eagerly exploited by the weevil.

This state of affairs could not last. If our own fields do not contain the insect amateur of the haricot the New World knows it well enough. By the road of commercial exchange, sooner or later some worm-eaten sack of haricots must bring it to Europe. The invasion is inevitable.

According to documents now before me, indeed, it has already taken place. Three or four years ago I received from Maillane, in the Bouches-du-Rhone, what I sought in vain in my own neighbourhood, although I questioned many a farmer and housewife, and astonished them by my questions. No one had ever seen the pest of the haricot; no one had ever heard of it. Friends who knew of my inquiries sent me from Maillane, as I have said, information that gave great satisfaction to my naturalist's curiosity. It was accompanied by a measure of haricots which were utterly and outrageously spoiled; every bean was riddled with holes, changed into a kind of sponge. Within them swarmed innumerable weevils, which recalled, by their diminutive size, the lentil-weevil, **Bruchus lenti**.

The senders told me of the loss experienced at Maillane. The odious little creature, they said, had destroyed the greater portion of the harvest. A veritable plague, such as had never before been known, had fallen upon the haricots, leaving the housewife barely a handful to put in the saucepan. Of the habits of the creature and its way of going to work nothing was known. It was for me to discover them by means of experiment.

Quick, then, let us experiment! The circumstances favour me. We are in the middle of June, and in my garden there is a bed of early haricots; the black Belgian haricots, sown for use in the kitchen. Since I must sacrifice the toothsome vegetable, let us loose the terrible destroyer on the mass of verdure. The development of the plant is at the requisite stage, if I may go by what the **Bruchus pisi** has already taught me; the flowers are abundant, and the pods are equally so; still green, and of all sizes.

I place on a plate two or three handfuls of the infested haricots, and set the populous heap in the full sunlight by the edge of my bed of beans. I can imagine what will happen. Those insects which are already free, and those which the stimulus of the sunshine will presently liberate, will emerge and take to their wings. Finding the maternal haricot close at hand they will take possession of the vines. I shall see them exploring pods and flowers, and before very long they will lay their eggs. That is how the pea-weevil would behave under similar conditions.

But no: to my surprise and confusion, matters do not fall out as I foresaw. For a few minutes the insects bustle about in the sunlight, opening and closing their wing-covers to ease the mechanism of flight; then one by one they fly away, mounting in the luminous air; they grow smaller and smaller to the sight, and are quickly lost to view. My persevering attentions have not met with the slightest success; not one of the weevils has settled on my haricots.

When the joys of liberty have been tasted will they return--to-night, to-morrow, or later? No, they do not return. All that week, at favourable hours, I inspect the rows of beans pod by pod, flower by flower; but never a Bruchus do I see, nor even an egg. Yet the season is propitious, for at this very moment the mothers imprisoned in my jars lay a profusion of eggs upon the dry haricots.

Next season I try again. I have at my disposal two other beds, which I have sown with the late haricot, the red haricot; partly for the use of the household, but principally for the benefit of the weevil. Arranged in convenient rows, the two crops will be ready, one in August and one in September or later.

With the red haricot I repeat the experiment already essayed with the black haricot. On several occasions, in suitable weather, I release large numbers of weevils from my glass jars, the general headquarters of the tribe. On each occasion the result is plainly negative. All through the season, until both crops are exhausted, I repeat my search almost daily; but I can never discover a single pod infested, nor even a single weevil perching on leaf or flower.

Certainly the inspection has not been at fault. The household is warned to respect certain rows of beans which I have reserved for myself. It is also requested to keep a look-out for eggs on all the pods gathered. I myself examine with a magnifying-glass all the haricots coming from my own or from neighbouring gardens before handing them over to the housewife to be shelled. All my trouble is wasted: there is not an egg to be seen.

To these experiments in the open air I add others performed under glass. I place, in some tall, narrow bottles, fresh haricot pods hanging from their stems; some green, others mottled with crimson, and containing seeds not far from mature. Each bottle is finally given a population of weevils. This time I obtain some eggs, but I am no further advanced; they are laid on the sides of the bottles, but not on the pods. Nevertheless, they hatch. For a few days I see the grubs wandering about, exploring the pods and the glass with equal zeal. Finally one and all perish without touching the food provided.

The conclusion to be drawn from these facts is obvious: the young and tender haricot is not the proper diet. Unlike the ***Bruchus pisi***, the female of the haricot-weevil refuses to trust her family to beans that are not hardened by age and desiccation; she refused to settle on my bean-patch because the food she required was not to be found there. What does she require? Evidently the mature, dry, hard haricot, which falls to earth with the sound of a small pebble. I hasten to satisfy her. I place in the bottles some very mature, horny pods, thoroughly desiccated by exposure to the sun. This time the family prospers, the grubs perforate the dry shell, reach the beans, penetrate them, and henceforth all goes well.

To judge by appearances, then, the weevil invades the granary. The beans are left standing in the fields until both plants and pods, shrivelled by the sun, are completely desiccated. The process of beating the pods to loosen and separate the beans is thus greatly facilitated. It is then that the weevil, finding matters to suit her, commences to lay her eggs. By storing his crop a little late the peasant stores the pest as well.

But the weevil more especially attacks the haricot when warehoused. Like the Calander-beetle, which nibbles the wheat in our granaries but despises the cereal while still on the stalk, it abhors the bean while tender, and prefers to establish itself in the peace and darkness of the storehouse. It is a formidable enemy to the merchant rather than to the peasant.

What a fury of destruction once the ravager is installed in the vegetable treasure-house! My bottles give abundant evidence of this. One single haricot bean shelters a numerous family; often as many as twenty members. And not one generation only exploits the bean, but three or four in the year. So long as the skin of the bean contains any edible matter, so long do new consumers establish themselves within it, so that the haricot finally becomes a mere shell stuffed with excreta. The skin, despised by the grubs, is a mere sac, pierced with holes as many as the inhabitants that have deserted it; the ruin is complete.

The ***Bruchus pisi***, a solitary hermit, consumes only so much of the pea as will leave a cell for the nymph; the rest remains intact, so that the pea may be sown, or it will even serve as food, if we can overcome our repugnance. The American insect knows nothing of these limitations; it empties the haricot completely and leaves a skinful of filth that I have seen the pigs refuse. America is anything but considerate when she sends us her entomological pests. We owe the Phylloxera to America; the Phylloxera, that calamitous insect against which our vine-growers wage incessant war: and to-day she is sending us the haricot-weevil, which threatens to be a plague of the future. A few experiments gave me some idea of the peril of such an invasion.

For nearly three years there have stood, on my laboratory table, some dozens of jars and bottles covered with pieces of gauze which prevent escape while permitting of a constant ventilation. These are the cages of my menagerie. In them I rear the haricot-weevil, varying the system of education at will. Amongst other things I have learned that this insect, far from being exclusive in its choice, will accommodate itself to most of our leguminous foods.

All the haricots suit it, black and white, red and variegated, large and small; those of the latest crop and those which have been many years in stock and are almost completely refractory to boiling water. The loose beans are attacked by preference, as being easier to invade, but when the loose beans are not available those in the natural shelter of their pods are attacked with equal zest. However dry and parchment-like the pods, the grubs have no difficulty in attaining the seeds. When attacked in the field or garden, the bean is attacked in this way through the pod. The bean known in Provence as the blind haricot--lou faiou borgne--a bean with a long pod, which is marked with a black spot at the navel, which has the look of a closed and blackened eye, is also greatly appreciated; indeed, I fancy my little guests show an obvious preference for this particular bean.

So far, nothing abnormal; the Bruchus does not wander beyond the limits of the botanical family ***Phaseolus***. But here is a characteristic that increases the peril, and shows us this lover of beans in an unexpected light. Without the slightest hesitation it accepts the dry pea, the bean, the vetch, the tare, and the chick-pea; it goes from one to the other, always satisfied; its offspring live and prosper in all these seeds as well as in the haricot. Only the lentil is refused, perhaps on account of its insufficient volume. The American weevil is a formidable experimentalist.

The peril would be much greater did the insect pass from leguminous seeds to cereals, as at first I feared it might. But it does not do so; imprisoned in my bottles together with a handful of wheat, barley, rice, or maize, the Bruchus invariably perished and left no offspring. The result was the same with oleaginous seeds: such as castor-oil and sunflower. Nothing outside the bean family is of any use to the Bruchus. Thus limited, its portion is none the less considerable, and it uses and abuses it with the utmost energy. The eggs are white, slender, and cylindrical. There is no method in their distribution, no choice in their deposition. The mother lays them singly or in little groups, on the walls of the jar as well as on the haricots. In her negligence she will even lay them on maize, coffee, castor-oil seeds, and other seeds, on which the newly born grubs will promptly perish, not finding them to their taste. What place has maternal foresight here? Abandoned no matter where in the heap of seeds, the eggs are always in place, as it is left to the grub to search and to find the points of invasion.

In five days at most the egg is hatched. A little white creature with a red-brown head emerges. It is a mere speck of a creature, just visible to the naked eye. Its body is thickened forward, to give more strength to its implements--its mandibles--which have to perforate the hard substance of the dry bean, which is as tough as wood. The larvae of the Buprestis and the Capricornis, which burrow in the trunks of trees, are similarly shaped. Directly it issues from the egg the wriggling creature makes off at random with an activity we should hardly expect in one so young. It wanders hither and thither, eager to find food and shelter as soon as possible.

Within twenty-four hours it has usually attained both. I see the tiny grub perforate the horny skin that covers the cotyledons; I watch its efforts; I surprise it sunk half-way in the commencement of a burrow, at the mouth of which is a white floury powder, the waste from the mandibles. It works its way inward and buries itself in the heart of the seed. It will emerge in the adult form in the course of about five weeks, so rapid is its evolution.

This hasty development allows of several generations in the year. I have recorded four. On the other hand, one isolated couple has furnished me with a family of eighty. Consider only the half of this number--supposing the sexes to be equal in number--and at the end of a year the couples issued from this original pair would be represented by the fortieth power of forty; in larvae they would represent the frightful total of more than five millions. What a mountain of haricots would be ravaged by such a legion!

The industry of the larvae reminds us at every point what we have learned from the *Bruchus pisi*. Each grub excavates a lodging in the mass of the bean, respecting the epidermis, and preparing a circular trap-door which the adult can easily open with a push at the moment of emergence. At the termination of the larval phase the lodgements are betrayed on the surface of the bean by so many shadowy circles. Finally the lid falls, the insect leaves its cell, and the haricot remains pierced by as many holes as it has nourished grubs.

Extremely frugal, satisfied with a little farinaceous powder, the adults seem by no means anxious to abandon the native heap or bin so long as there are beans untouched. They mate in the interstices of the heap; the mothers sow their eggs at random; the young larvae establish themselves some in beans that are so far intact, some in beans which are perforated but not yet exhausted; and all through the summer the operations of breeding are repeated once in every five weeks. The last generation of the year--that of September or October--sleeps in its cells until the warm weather returns.

If the haricot pest were ever to threaten us seriously it would not be very difficult to wage a war of extermination against it. Its habits teach us what tactics we ought to follow. It exploits the dried and gathered crop in the granary or the storehouse. If it is difficult to attack it in the open it would also be useless. The greater part of its affairs are managed elsewhere, in our storehouses. The enemy establishes itself under our roof and is ready to our hand. By means of insecticides defence should be relatively easy.

CHAPTER XX
THE GREY LOCUST

I have just witnessed a moving spectacle: the last moult of a locust; the emergence of the adult from its larval envelope. It was magnificent. I am speaking of the Grey Locust, the colossus among our acridians,[10] which is often seen among the vines in September when the grapes are gathered. By its size--and it grows as long as a man's finger--it lends itself to observation better than any other of its tribe.

The larva, disgustingly fat, like a rude sketch of the perfect insect, is commonly of a tender green; but it is sometimes of a bluish green, a dirty yellow, or a ruddy brown, or even an ashen grey, like the grey of the adult cricket. The corselet is strongly keeled and indented, and is sprinkled with fine white spots. As powerful as in the adult insect, the hind-leg has a corpulent haunch, streaked with red, and a long shin like a two-edged saw.

The elytra, which in a few days will extend far beyond the tip of the abdomen, are at present too small triangular wing-like appendages, touching along their upper edges, and continuing and emphasising the keel or ridge of the corselet. Their free ends stick up like the gable of a house. They remind one of the skirts of a coat, the maker of which has been ludicrously stingy with the cloth, as they merely cover the creature's nakedness at the small of the back. Underneath there are two narrow appendages, the germs of the wings, which are even smaller than the elytra. The sumptuous, elegant sails of to-morrow are now mere rags, so miserly in their dimensions as to be absolutely grotesque. What will emerge from these miserable coverings? A miracle of grace and amplitude.

Let us observe the whole process in detail. Feeling itself ripe for transformation, the insect climbs up the wire-gauze cover by means of its hinder and intermediate limbs. The fore-limbs are folded and crossed on the breast, and are not employed in supporting the insect, which hangs in a reversed position, the back downwards. The triangular winglets, the sheaths of the elytra, open along their line of juncture and separate laterally; the two narrow blades, which contain the wings, rise in the centre of the interval and slightly diverge. The proper position for the process of moulting has now been assumed and the proper stability assured.

The first thing to do is to burst the old skin. Behind the corselet, under the pointed roof of the prothorax, a series of pulsations is produced by alternate inflation and deflation. A similar state of affairs is visible in front of the neck, and probably under the entire surface of the yielding carapace. The fineness of the membrane at the articulations enables us to perceive it at these unarmoured points, but the cuirass of the corselet conceals it in the central portion.

At these points the circulatory reserves of the insect are pulsing in tidal onsets. Their gradual increase is betrayed by pulsations like those of a hydraulic ram. Distended by this rush of humours, by this injection in which the organism concentrates all its forces, the outer skin finally splits along the line of least resistance which the subtle previsions of life have prepared. The fissure extends the whole length of the corselet, opening precisely along the ridge of the keel, as though the two symmetrical halves had been soldered together. Unbreakable elsewhere, the envelope has yielded at this median point, which had remained weaker than the rest of the sheath. The fissure runs back a little way until it reaches a point between the attachments of the wings; on the head it runs forward as far as the base of the antennae, when it sends a short ramification right and left.

Through this breach the back is seen; quite soft, and very pale, with scarcely a tinge of grey. Slowly it curves upwards and becomes more and more strongly hunched; at last it is free.

The head follows, withdrawing itself from its mask, which remains in place, intact in the smallest detail, but looking very strange with its great unseeing glassy eyes. The sheaths of the antennae, without a wrinkle, without the least derangement, and in their natural place, hang over this dead, translucid face.

In emerging from their narrow sheaths, which clasped them so tightly and precisely, the thread-like antennae have evidently met with no resistance, or the sheaths would have been turned inside out, or crumpled out of shape, or wrinkled at least. Without harming the jointed or knotted covers, the contents, of equal volume and equally knotty, have slipped out as easily as though they were smooth, slippery objects sliding out of a loose sheath. The method of extraction is still more astonishing in the case of the hind-legs.

It is now, however, the turn of the front and intermediate pairs of legs. They pull out of their gauntlets and leggings without the least hitch; nothing is torn, nothing buckled; the outer skin is not even crumpled, and all the tissues remain in their natural position. The insect is now hanging from the dome of the cover solely by the claws of the long hind-legs. It hangs in an almost vertical position, the head downwards, swinging like a pendulum if I touch the cover. Four tiny, steely claws are its only support. If they gave or unclasped themselves the insect would be lost, as it is as yet unable to unfurl its enormous wings; but even had the wings emerged they could not grip the air in time to save the creature from the consequences of a fall. But the four claws hold fast; life, before withdrawing from them, left them rigidly contracted, so that they should support without yielding the struggles and withdrawals to follow.

Now the wing-covers and wings emerge. These are four narrow strips, vaguely seamed and furrowed, like strings of rolled tissue-paper. They are barely a quarter of their final length.

They are so soft that they bend under their own weight, and hang down the creature's sides in the reverse of their normal position. The free extremities, which normally point backwards, are now pointing towards the cricket's head as it hangs reversed. The organs of future flight are like four leaves of withered foliage shattered by a terrific rainstorm.

A profound transformation is necessary to bring the wings to their final perfection. The inner changes are already at work; liquids are solidifying; albuminous secretions are bringing order out of chaos; but so far no outward sign betrays what is happening in the mysterious laboratory of the organism. All seems inert and lifeless.

In the meantime the posterior limbs disengage themselves. The great haunches become visible, streaked on the inner faces with a pale rose, which rapidly turns to a vivid crimson. Emergence is easy, the thick and muscular upper portion of the haunch preparing the way for the narrower part of the limb.

It is otherwise with the shank. This, in the adult insect, is armed along its whole length by a double series of stiff, steely spines. Moreover, the lower extremity is terminated by four strong spurs. The shank forms a veritable saw, but with two parallel sets of teeth; and it is so strongly made that it may well be compared, the question of size apart, to the great saw of a quarry-man.

The shank of the larva has the same structure, so that the object to be extracted is enclosed in a scabbard as awkwardly shaped as itself. Each spur is enclosed in a similar spur; each tooth engages in the hollow of a similar tooth, and the sheath is so closely moulded upon the shank that a no more intimate contact could be obtained by replacing the envelope by a layer of varnish applied with a brush.

Nevertheless the tibia, long and narrow as it is, issues from its sheath without catching or sticking anywhere. If I had not repeatedly seen the operation I could not believe it possible; for the discarded sheath is absolutely intact from end to end. Neither the terminal spurs nor the double rows of spines do the slightest damage to the delicate mould. The long-toothed saw leaves the delicate sheath unbroken, although a puff of the breath is enough to tear it; the ferocious spurs slip out of it without leaving so much as a scratch.

I was far from expecting such a result. Having the spiny weapons of the legs in mind, I imagined that those limbs would moult in scales and patches, or that the sheathing would rub off like a dead scarf-skin. How completely the reality surpassed my anticipations!

From the spurs and spines of the sheath, which is as thin as the finest gold-beaters' skin, the spurs and spines of the leg, which make it a most formidable weapon, capable of cutting a piece of soft wood, emerge without the slightest display of violence, without a hitch of any kind; and the empty skin remains in place. Still clinging by its claws to the top of the wire cover, it is untorn, unwrinkled, uncreased. Even the magnifying-glass fails to show a trace of rough usage. Such as the skin was before the cricket left it, so it is now. The legging of dead skin remains in its smallest details the exact replica of the living limb.

If any one asked you to extract a saw from a scabbard exactly moulded upon the steel, and to conduct the operation without the slightest degree of tearing or scratching, you would laugh at the flagrant impossibility of the task. But life makes light of such absurdities; it has its methods of performing the impossible when such methods are required. The leg of the locust affords us such an instance.

Hard as it is when once free of its sheath, the serrated tibia would absolutely refuse to leave the latter, so closely does it fit, unless it were torn to pieces. Yet the difficulty must be evaded, for it is indispensable that the sheaths of the legs should remain intact, in order to afford a firm support until the insect is completely extricated.

The leg in process of liberation is not the leg with which the locust makes its leaps; it has not as yet the rigidity which it will soon acquire. It is soft, and eminently flexible. In those portions which the progress of the moult exposes to view I see the legs bend under the mere weight of the suspended insect when I tilt the supporting cover. They are as flexible as two strips of elastic indiarubber. Yet even now consolidation is progressing, for in a few minutes the proper rigidity will be acquired.

Further along the limbs, in the portions which the sheathing still conceals, the legs are certainly softer still, and in the state of exquisite plasticity--I had almost said fluidity--which allows them to pass through narrow passages almost as a liquid flows.

The teeth of the saws are already there, but have nothing of their imminent rigidity. With the point of a pen-knife I can partially uncover a leg and extract the spines from their serrated mould. They are germs of spines; flexible buds which bend under the slightest pressure and resume their position the moment the pressure is removed.

These needles point backwards as the leg is drawn out of the sheath; but they re-erect themselves and solidify as they emerge. I am witnessing not the mere removal of leggings from limbs already clad in finished armour, but a kind of creation which amazes one by its promptitude.

Very much in the same way, but with far less delicate precision, the claws of the crayfish, at the period of the moult, withdraw the soft flesh of their double fingers from their stony sheath.

Finally the long stilt-like legs are free. They are folded gently against the furrowed thighs, thus to mature undisturbed. The abdomen begins to emerge. Its fine tunic-like covering splits, and wrinkles, but still encloses the extremity of the abdomen, which adheres to the moulted skin for some little time longer. With the exception of this one point the entire insect is now uncovered.

It hangs head downwards, like a pendulum, supported by the talons of the now empty leg-cases. During the whole of the lengthy and meticulous process the four talons have never yielded. The whole operation has been conducted with the utmost delicacy and prudence.

The insect hangs motionless, held by the tip of the abdomen. The abdomen is disproportionately distended; swollen, apparently, by the reserve of organisable humours which the expansion of the wings and wing-covers will presently employ. Meanwhile the creature rests and recovers from its exertions. Twenty minutes of waiting elapse.

Then, exerting the muscles of the back, the suspended insect raises itself and fixes the talons of the anterior limbs in the empty skin above it. Never did acrobat, hanging by the toes to the bar of a trapeze, raise himself with so stupendous a display of strength in the loins. This gymnastic feat accomplished, the rest is easy.

With the purchase thus obtained the insect rises a little and reaches the wire gauze, the equivalent of the twig which would be chosen for the site of the transformation in the open fields. It holds to this with the four anterior limbs. Then the tip of the abdomen is finally liberated, and suddenly, shaken by the final struggle, the empty skin falls to the ground.

This fall is interesting, and reminds me of the persistence with which the empty husk of the Cigale braves the winds of winter, without falling from its supporting twig. The transfiguration of the locust takes place very much as does that of the Cigale. How is it then that the acridian trusts to a hold so easily broken?

The talons of the skin hold firmly so long as the labour of escape continues, although one would expect it to shake the firmest grip; yet they yield at the slightest shock when the labour is terminated. There is evidently a condition of highly unstable equilibrium; showing once more with what delicate precision the insect escapes from its sheath.

For want of a better term I said "escape." But the word is ill chosen; for it implies a certain amount of violence, and no violence must be employed, on account of the instability of equilibrium already mentioned. If the insect, shaken by a sudden effort, were to lose its hold, it would be all up with it. It would slowly shrivel on the spot; or at best its wings, unable to expand, would remain as miserable scraps of tissue. The locust does not tear itself away from its sheath; it delicately insinuates itself out of it--I had almost said flows. It is as though it were expelled by a gentle pressure.

Let us return to the wings and elytra, which have made no apparent progress since their emergence from their sheaths. They are still mere stumps, with fine longitudinal seams; almost like little ropes'-ends. Their expansion, which will occupy more than three hours, is reserved for the end, when the insect is completely moulted and in its normal position.

We have just seen the insect turn head uppermost. This reversal causes the wings and elytra to fall into their natural position. Extremely flexible, and yielding to their own weight, they had previously drooped backwards with their free extremities pointing towards the head of the insect as it hung reversed.

Now, still by reason of their own weight, their position is rectified and they point in the normal direction. They are no longer curved like the petals of a flower; they no longer point the wrong way; but they retain the same miserable aspect.

In its perfect state the wing is like a fan. A radiating bundle of strong nervures runs through it in the direction of its length and forms the framework of the fan, which is readily furled and unfurled. The intervals are crossed by innumerable cross-nervures of slighter substance, which make of the whole a network of rectangular meshes. The elytrum, which is heavier and much less extensive, repeats this structure.

At present nothing of this mesh-work is visible. Nothing can be seen but a few wrinkles, a few flexuous furrows, which announce that the stumps are bundles of tissue cunningly folded and reduced to the smallest possible volume.

The expansion of the wing begins near the shoulder. Where nothing precise could be distinguished at the outset we soon perceive a diaphanous surface subdivided into meshes of beautiful precision.

Little by little, with a deliberation that escapes the magnifier, this area increases its bounds, at the expense of the shapeless bundle at the end of the wing. In vain I let my eyes rest on the spot where the expanding network meets the still shapeless bundle; I can distinguish nothing. But wait a little, and the fine-meshed tissues will appear with perfect distinctness.

To judge from this first examination, one would guess that an organisable fluid is rapidly congealing into a network of nervures; one seems to be watching a process of crystallisation comparable, in its rapidity, to that of a saturated saline solution as seen through a microscope. But no; this is not what is actually happening. Life does not do its work so abruptly.

I detach a half-developed wing and bring it under the powerful eye of the microscope. This time I am satisfied. On the confines of the transparent network, where an extension of that network seems to be gradually weaving itself out of nothing, I can see that the meshes are really already in existence. I can plainly recognise the longitudinal nervures, which are already stiff; and I can also see--pale, and without relief--the transverse nervures. I find them all in the terminal stump, and am able to spread out a few of its folds under the microscope.

It is obvious that the wing is not a tissue in the process of making, through which the procreative energy of the vital juices is shooting its shuttle; it is a tissue already complete. To be perfect it lacks only expansion and rigidity, just as a piece of lace or linen needs only to be ironed.

In three hours or more the explanation is complete. The wings and elytra stand erect over the locust's back like an immense set of sails; at first colourless, then of a tender green, like the freshly expanded wings of the Cigale. I am amazed at their expanse when I think of the miserable stumps from which they have expanded. How did so much material contrive to occupy so little space?

There is a story of a grain of hemp-seed that contained all the body-linen of a princess. Here we have something even more astonishing. The hemp-seed of the story needed long years to germinate, to multiply, and at last to give the quantity of hemp required for the trousseau of a princess; but the germ of the locust's wing has expanded to a magnificent sail in a few short hours.

Slowly the superb erection composed of the four flat fan-like pinions assumes rigidity and colour. By to-morrow the colour will have attained the requisite shade. For the first time the wings close fan-wise and lie down in their places; the elytra bend over at their outer edges, forming a flange which lies snugly over the flanks. The transformation is complete. Now the great locust has only to harden its tissues a little longer and to tan the grey of its costume in the ecstasy of the sunshine. Let us leave it to its happiness, and return to an earlier moment.

The four stumps which emerge from their coverings shortly after the rupture of the corselet along its median line contain, as we have seen, the wings and elytra with their innumerable nervures. If not perfect, at least the general plan is complete, with all its innumerable details. To expand these miserable bundles and convert them into an ample set of sails it is enough that the organism, acting like a force-pump, should force into the channels already prepared a stream of humours kept in reserve for this moment and this purpose, the most laborious of the whole process. As the capillary channels are prepared in advance a slight injection of fluid is sufficient to cause expansion.

But what were these four bundles of tissue while still enclosed in their sheaths? Are the wing-sheaths and the triangular winglets of the larva the moulds whose folds, wrinkles, and sinuosities form their contents in their own image, and so weave the network of the future wings and wing-covers?

Were they really moulds we might for a moment be satisfied. We might tell ourselves: It is quite a simple matter that the thing moulded should conform to the cavity of the mould. But the simplicity is only apparent, for the mould in its turn must somewhere derive the requisite and inextricable complexity. We need not go so far back; we should only be in darkness. Let us keep to the observable facts.

I examine with a magnifying-glass one of the triangular coat-tails of a larva on the point of transformation. I see a bundle of moderately strong nervures radiating fan-wise. I see other nervures in the intervals, pale and very fine. Finally, still more delicate, and running transversely, a number of very short nervures complete the pattern.

Certainly this resembles a rough sketch of the future wing-case; but how different from the mature structure! The disposition of the radiating nervures, the skeleton of the structure, is not at all the same; the network formed by the cross-nervures gives no idea whatever of the complex final arrangement. The rudimentary is succeeded by the infinitely complex; the clumsy by the infinitely perfect, and the same is true of the sheath of the wing and the final condition of its contents, the perfect wing.

It is perfectly evident, when we have the preparatory as well as the final condition of the wing before our eyes, that the wing-sheath of the larva is not a simple mould which elaborates the tissue enclosed in its own image and fashions the wing after the complexities of its own cavity.

The future wing is not contained in the sheath as a bundle, which will astonish us, when expanded, by the extent and extreme complication of its surface. Or, to speak more exactly, it is there, but in a potential state. Before becoming an actual thing it is a virtual thing which is not yet, but is capable of becoming. It is there as the oak is inside the acorn.

A fine transparent cushion limits the free edge of the embryo wing and the embryo wing-case. Under a powerful microscope we can perceive therein a few doubtful lineaments of the future lace-work. This might well be the factory in which life will shortly set its materials in movement. Nothing more is visible; nothing that will make us foresee the prodigious network in which each mesh must have its form and place predetermined with geometrical exactitude.

In order that the organisable material can shape itself as a sheet of gauze and describe the inextricable labyrinth of the nervuration, there must be something better and more wonderful than a mould. There is a prototypical plan, an ideal pattern, which imposes a precise position upon each atom of the tissue. Before the material commences to circulate the configuration is already virtually traced, the courses of the plastic currents are already mapped out. The stones of our buildings co-ordinate according to the considered plan of the architect; they form an ideal assemblage before they exist as a concrete assemblage.

Similarly, the wing of a cricket, that wonderful piece of lace-work emerging from a tiny sheath, speaks to us of another Architect, the author of the plans according to which life labours.

The genesis of living creatures offers to our contemplation an infinity of wonders far greater than this matter of a locust's wing; but in general they pass unperceived, obscured as they are by the veil of time.

Time, in the deliberation of mysteries, deprives us of the most astonishing of spectacles except our spirits be endowed with a tenacious patience. Here by exception the fact is accomplished with a swiftness that forces the attention.

Whosoever would gain, without wearisome delays, a glimpse of the inconceivable dexterity with which the forces of life can labour, has only to consider the great locust of the vineyard. The insect will show him that which is hidden from our curiosity by extreme deliberation in the germinating seed, the opening leaf, and the budding flower. We cannot see the grass grow; but we can watch the growth of the locust's wings.

Amazement seizes upon us before this sublime phantasmagoria of the grain of hemp which in a few hours has been transmuted into the finest cloth. What a mighty artist is Life, shooting her shuttle to weave the wings of the locust--one of those insignificant insects of whom long ago Pliny said: ***In his tam parcis, fere nullis, quae vis, quae sapientia, quam inextricabilis perfectio!***

How truly was the old naturalist inspired! Let us repeat with him: "What power, what wisdom, what inconceivable perfection in this least of secrets that the vineyard locust has shown us!"

I have heard that a learned inquirer, for whom life is only a conflict of physical and chemical forces, does not despair of one day obtaining artificially organisable matter--protoplasm, as the official jargon has it. If it were in my power I should hasten to satisfy this ambitious gentleman.

But so be it: you have really prepared protoplasm. By force of meditation, profound study, minute care, impregnable patience, your desire is realised: you have extracted from your apparatus an albuminous slime, easily corruptible and stinking like the devil at the end of a few days: in short, a nastiness. What are you going to do with it?

Organise something? Will you give it the structure of a living edifice? Will you inject it with a hypodermic syringe between two impalpable plates to obtain were it only the wing of a fly?

That is very much what the locust does. It injects its protoplasm between the two surfaces of an embryo organ, and the material forms a wing-cover, because it finds as guide the ideal archetype of which I spoke but now. It is controlled in the labyrinth of its course by a device anterior to the injection: anterior to the material itself.

This archetype, the co-ordinator of forms; this primordial regulator; have you got it on the end of your syringe? No! Then throw away your product. Life will never spring from that chemical filth.

CHAPTER XXI
THE PINE-CHAFER

The orthodox denomination of this insect is **Melolontha fullo**, Lin. It does not answer, I am very well aware, to be difficult in matters of nomenclature; make a noise of some sort, affix a Latin termination, and you will have, as far as euphony goes, the equivalent of many of the tickets pasted in the entomologist's specimen boxes. The cacophony would be excusable if the barbarous term signified nothing but the creature signified; but as a rule this name possesses, hidden in its Greek or other roots, a certain meaning in which the novice hopes to find instruction.

The hope is a delusion. The learned term refers to subtleties difficult to comprehend, and of very indifferent importance. Too often it leads the student astray, giving him glimpses that have nothing whatever in common with the truth as we know it from observation. Very often the errors implied by such names are flagrant; sometimes the allusions are ridiculous, grotesque, or merely imbecile. So long as they have a decent sound, how infinitely preferable are locutions in which etymology finds nothing to dissect! Of such would be the word *fullo*, were it not that it already has a meaning which immediately occurs to the mind. This Latin expression means a *fuller*; a person who kneads and presses cloth under a stream of water, making it flexible and ridding it of the asperities of weaving. What connection has the subject of this chapter with the fuller of cloth? I may puzzle my head in vain: no acceptable reply will occur to me.

The term *fullo* as applied to an insect is found in Pliny. In one chapter the great naturalist treats of remedies against jaundice, fevers, and dropsy. A little of everything enters into this antique pharmacy: the longest tooth of a black dog; the nose of a mouse wrapped in a pink cloth; the right eye of a green lizard torn from the living animal and placed in a bag of kid-skin; the heart of a serpent, cut out with the left hand; the four articulations of the tail of a scorpion, including the dart, wrapped tightly in a black cloth, so that for three days the sick man can see neither the remedy nor him that applies it; and a number of other extravagances. We may well close the book, alarmed at the slough of the imbecility whence the art of healing has come down to us.

In the midst of these imbecilities, the preludes of medicine, we find a mention of the "fuller." ***Tertium qui vocatur fullo, albis guttis, dissectum utrique lacerto adalligant***, says the text. To treat fevers divide the fuller beetle in two parts and apply half under the right arm and half under the left.

Now what did the ancient naturalist mean by the term "fuller beetle"? We do not precisely know. The qualification ***albis guttis***, white spots, would fit the Pinechafer well enough, but it is not sufficient to make us certain. Pliny himself does not seem to have been very certain of the identity of the remedy. In his time men's eyes had not yet learned to see the insect world. Insects were too small; they were well enough for amusing children, who would tie them to the end of a long thread and make them walk in circles, but they were not worthy of occupying the attention of a self-respecting man.

Pliny apparently derived the word from the country-folk, always poor observers and inclined to extravagant denominations. The scholar accepted the rural locution, the work perhaps of the imagination of childhood, and applied it at hazard without informing himself more particularly. The word came down to us embalmed with age; our modern naturalists have accepted it, and thus one of our handsomest insects has become the "fuller." The majesty of antiquity has consecrated the strange appellation.

In spite of all my respect for the antique, I cannot myself accept the term "fuller," because under the circumstances it is absurd. Common sense should be considered before the aberrations of nomenclature. Why not call our subject the Pinechafer, in reference to the beloved tree, the paradise of the insect during the two or three weeks of its aerial life? Nothing could be simpler, or more appropriate, to give the better reason last.

We have to wander for ages in the night of absurdity before we reach the radiant light of the truth. All our sciences witness to this fact; even the science of numbers. Try to add a column of Roman figures; you will abandon the task, stupefied by the confusion of symbols; and will recognise what a revolution was made in arithmetic by the discovery of the zero. Like the egg of Columbus, it was a very little thing, but it had to be thought of.

While hoping that the future will sink the unfortunate "fuller" in oblivion, we will use the term "pine chafer" between ourselves. Under that name no one can possibly mistake the insect in question, which frequents the pine-tree only.

It has a handsome and dignified appearance, rivalling that of **Oryctes nasicornis**. Its costume, if it has not the metallic splendour dear to the Scarabaei, the Buprestes and the rose-beetles, is at least unusually elegant. A black or chestnut background is thickly sown with capriciously shaped spots of white velvet; a fashion both modest and handsome.

The male bears at the end of his short antennae a kind of plume consisting of seven large superimposed plates or leaves, which, opening and closing like the sticks of a fan, betray the emotions that possess him. At first sight it seems that this magnificent foliage must form a sense-organ of great perfection, capable of perceiving subtle odours, or almost inaudible vibrations of the air, or other phenomena to which our senses fail to respond; but the female warns us that we must not place too much reliance on such ideas; for although her maternal duties demand a degree of impressionability at least as great as that of the male, yet the plumes of her antennae are extremely meagre, containing only six narrow leaves.

What then is the use of the enormous fan-like structure of the male antennae? The seven-leaved apparatus is for the Pine-chafer what his long vibrating horns are to the Cerambyx and the panoply of the head to the Onthophagus and the forked antlers of the mandibles to the Stag-beetle. Each decks himself after his manner in these nuptial extravagances.

This handsome chafer appears towards the summer solstice, almost simultaneously with the first Cigales. The punctuality of its appearance gives it a place in the entomological calendar, which is no less punctual than that of the seasons. When the longest days come, those days which seem endless and gild the harvests, it never fails to hasten to its tree. The fires of St. John, reminiscences of the festivals of the Sun, which the children light in the village streets, are not more punctual in their date.

At this season, in the hours of twilight, the Pine-chafer comes every evening if the weather is fine, to visit the pine-trees in the garden. I follow its evolutions with my eyes. With a silent flight, not without spirit, the males especially wheel and wheel about, extending their great antennary plumes; they go to and fro, to and fro, a procession of flying shadows upon the pale blue of the sky in which the last light of day is dying. They settle, take flight again, and once more resume their busy rounds. What are they doing up there during the fortnight of their festival?

The answer is evident: they are courting their mates, and they continue to render their homage until the fall of night. In the morning both males and females commonly occupy the lower branches. They lie there isolated, motionless, indifferent to passing events. They do not avoid the hand about to seize them. Most of them are hanging by their hind legs and nibbling the pine-needles; they seem to be gently drowsing with the needles at their mouths. When twilight returns they resume their frolics.

To watch these frolics in the tops of the trees is hardly possible; let us try to observe them in captivity. Four pairs are collected in the morning and placed, with some twigs off the pine-tree, in a spacious; cage. The sight is hardly worth my attention; deprived of the possibility of flight, the insects cannot behave as in the open. At most I see a male from time to time approaching his beloved; he spreads out the leaves of his antennae, and agitates them so that they shiver slightly; he is perhaps informing himself if he is welcome. Thereupon he puts on his finest airs and exhibits his attainments. It is a useless display; the female is motionless, as though insensible to these demonstrations. Captivity has sorrows that are hard to overcome. This was all that I was able to see. Mating, it appears, must take place during the later hours of the night, so that I missed the propitious moment.

One detail in particular interested me. The Pine-chafer emits a musical note. The female is as gifted as the male. Does the lover make use of his faculty as a means of seduction and appeal? Does the female answer the chirp of her *innamorata* by a similar chirp? That this may be so under normal conditions, amidst the foliage of the pines, is extremely probable; but I can make no assertion, as I have never heard anything of the kind either among the pines or in my laboratory.

The sound is produced by the extremity of the abdomen, which gently rises and falls, rubbing, as it does so, with its last few segments, the hinder edge of the wing-covers, which are held firm and motionless. There is no special equipment on the rubbing surface nor on the surface rubbed. The magnifying-glass looks in vain for the fine striations usually found in the musical instruments of the insect world. All is smooth on either hand. How then is the sound engendered?

Rub the end of the moistened finger on a strip of glass, or a window-pane, and you will obtain a very audible sound, somewhat analogous to that emitted by the chafer. Better still, use a scrap of indiarubber to rub the glass with, and you will re-produce with some fidelity the sound in question. If the proper rhythm is observed the imitation is so successful that one might well be deceived by it.

In the musical apparatus of the Pine-chafer the pad of the finger-tip and the scrap of indiarubber are represented by the soft abdomen of the insect, and the glass is represented by the blade of the wing-cover, which forms a thin, rigid plate, easily set in vibration. The sound-mechanism of the Pine-chafer is thus of the very simplest description.

NOTES:

[1: Whether the Cigale is absolutely deaf or not, it is certain that one Cigale would be able to perceive another's cry. The vibrations of the male Cigale's cry would cause a resonance, a vibration, in the body cavities of other male Cigales, and to a lesser extent in the smaller cavities in the bodies of the females. Other sounds would cause a slight shock, if loud enough, but not a perceptible vibration May not this vibration--felt as in a cathedral we feel the vibrations of the organ-pipes in the bones of the chest and head or on the covers of the hymn-book in our hands--serve to keep the insects together, and enable the females to keep within sight of the males? The sight of an insect is in one sense poor--it consists of a kind of mosaic picture, and for one insect to distinguish another clearly the distance between them must not be very great. Certain gregarious birds and fish whose colouring is protective have a habit of showing their white bellies as they swerve on changing their direction. These signals help to keep the flock together. The white scut of the rabbit and of certain deer is a signal for other deer or rabbits to follow a frightened flock. It is obviously to the advantage of the Cigale to follow a gregarious habit, if only for purposes of propagation, for this would be facilitated by the sexes keeping together, and, deaf or otherwise, the vibrations of its cry would enable it to do so. It would be easy to show *a priori* that the perception of such vibrations must cause the insect pleasure, as they stimulate a nervous structure attuned to the perception or capable of the production of certain complex vibrations. The discord of the cry is caused by the fact that it consists of a number of vibrations of different pitch. Some would set the contents of the male resonating cavities in vibration; others would affect the less regular cavities in the thorax of the female. We might compare the Cigale's cry to a sheep-bell. That it is felt and not heard explains its loudness and its grating quality. A Cigale with the resonating cavities destroyed would possibly be lost. The experiment is worth trying.--[TRANS.]]

[2: It is not easy to understand why the Mantis should paralyse the cricket with terror while the latter will immediately escape when threatened by other enemies. As many species of Mantis exactly mimic sticks and leaves when motionless for purposes of defence, is it not possible that they mimic their surroundings for purposes of offence as well? It is easy and natural to say that the Mantis presents a terrifying aspect. It does to us, by association; but how can we say that it represents anything of the sort to the probably hypnotic or automatic consciousness of the cricket? What does it really represent, as seen from below? A twig, terminating in a bud, with two branching twigs growing from it, and a harmless nondescript fly or butterfly perched on the back of it. The combination of a familiar sight and a threatening sound would very plausibly result in cautious immobility. As for its instantaneous assumption of the pose, to move instantaneously is the next best thing to not moving at all. It is less likely to startle than a slow movement. Twigs which have been bent get suddenly released in the natural course of events; they do not move slowly. The instantaneous appearance of a twig where no twig was before may possibly give the victim pause; it may halt out of caution, not out of terror.--[TRANS.]]

[3: The word "butterfly" is here used, as is the French *papillon*, as a general term for all Lepidoptera; the insect in question is of course a moth.]

[4: Now classified as *Lasiocampa quercus*.--[TRANS.]]

[5: *Rabasso* is the Provencal name for the truffle; hence a truffle-hunter is known as a *rabassier*.]

[6: Since these lines were written I have found it consuming one of the true tuberaceae, the *Tuber Requienii*, Tul., of the size of a cherry.]

[7: The difficulty in conceiving this theory lies in the fact that the waves travel in straight lines. On the other hand, matter in a state of degradation may expel particles highly energised and of enormous velocity. Most antennae are covered with hairs of inconceivable fineness; others may contain cavities of almost infinite minuteness. Is it not thinkable that they are able to detect, in the gaseous atmosphere, floating particles that are not gaseous? This would not prevent the specialisation of antennae as mere feelers in some insects and crustaceans. The difficulty of such a supposition lies in the fact of discrimination; but if we did not possess a sense of taste or smell discrimination would seem inconceivable in their case also.--[TRANS.]]

[8: This classification is now superseded; the Pea and Bean Weevils--Bruchus pisi and *Bruchus lenti*--are classed as Bruchidae, in the series of Phytophaga. Most of the other weevils are classed as Curculionidae, series Rhyncophora.--[TRANS.]]

[9: The Christmas number (Noel) of the *Annales politiques et litteraires: Les*

Enfants juges par leurs peres, 1901.]

[10: The American usage is to call acridians grasshoppers and Locustidae locusts. The English usage is to call Locustidae grasshoppers and acridians locusts. The Biblical locust is an acridian.]

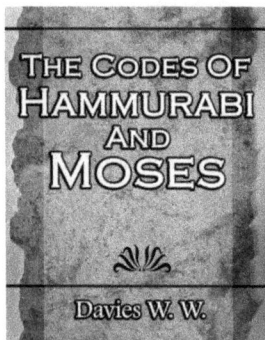

The Codes Of Hammurabi And Moses
W. W. Davies

QTY

The discovery of the Hammurabi Code is one of the greatest achievements of archaeology, and is of paramount interest, not only to the student of the Bible, but also to all those interested in ancient history...

Religion   ISBN: *1-59462-338-4*   **Pages:132**
MSRP $12.95

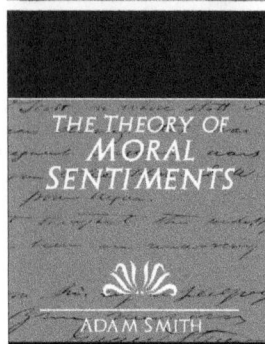

The Theory of Moral Sentiments
Adam Smith

QTY

This work from 1749. contains original theories of conscience amd moral judgment and it is the foundation for systemof morals.

Philosophy ISBN: *1-59462-777-0*   **Pages:536**
MSRP $19.95

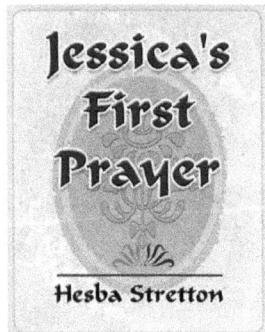

Jessica's First Prayer
Hesba Stretton

QTY

In a screened and secluded corner of one of the many railway-bridges which span the streets of London there could be seen a few years ago, from five o'clock every morning until half past eight, a tidily set-out coffee-stall, consisting of a trestle and board, upon which stood two large tin cans, with a small fire of charcoal burning under each so as to keep the coffee boiling during the early hours of the morning when the work-people were thronging into the city on their way to their daily toil...

Childrens   ISBN: *1-59462-373-2*   **Pages:84**
MSRP $9.95

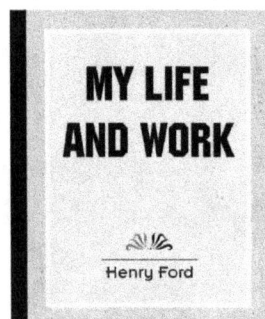

My Life and Work
Henry Ford

QTY

Henry Ford revolutionized the world with his implementation of mass production for the Model T automobile. Gain valuable business insight into his life and work with his own auto-biography... "We have only started on our development of our country we have not as yet, with all our talk of wonderful progress, done more than scratch the surface. The progress has been wonderful enough but..."

Biographies/   ISBN: *1-59462-198-5*   **Pages:300**
MSRP $21.95

www.bookjungle.com *email: sales@bookjungle.com fax: 630-214-0564 mail: Book Jungle PO Box 2226 Champaign, IL 61825*

QTY

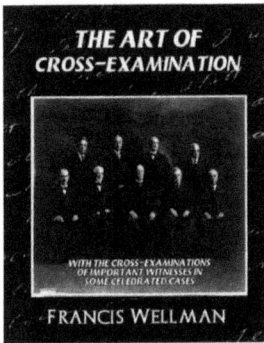

The Art of Cross-Examination
Francis Wellman

I presume it is the experience of every author, after his first book is published upon an important subject, to be almost overwhelmed with a wealth of ideas and illustrations which could readily have been included in his book, and which to his own mind, at least, seem to make a second edition inevitable. Such certainly was the case with me; and when the first edition had reached its sixth impression in five months, I rejoiced to learn that it seemed to my publishers that the book had met with a sufficiently favorable reception to justify a second and considerably enlarged edition. ..

Pages:412

Reference   ISBN: *1-59462-647-2*   *MSRP $19.95*

QTY

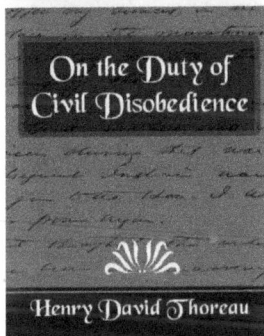

On the Duty of Civil Disobedience
Henry David Thoreau

Thoreau wrote his famous essay, On the Duty of Civil Disobedience, as a protest against an unjust but popular war and the immoral but popular institution of slave-owning. He did more than write—he declined to pay his taxes, and was hauled off to gaol in consequence. Who can say how much this refusal of his hastened the end of the war and of slavery ?

Law   ISBN: *1-59462-747-9*   **Pages:48**
MSRP $7.45

QTY

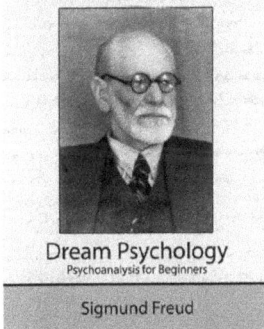

Dream Psychology Psychoanalysis for Beginners
Sigmund Freud

Sigmund Freud, born Sigismund Schlomo Freud (May 6, 1856 - September 23, 1939), was a Jewish-Austrian neurologist and psychiatrist who co-founded the psychoanalytic school of psychology. Freud is best known for his theories of the unconscious mind, especially involving the mechanism of repression; his redefinition of sexual desire as mobile and directed towards a wide variety of objects; and his therapeutic techniques, especially his understanding of transference in the therapeutic relationship and the presumed value of dreams as sources of insight into unconscious desires.

Pages:196

Psychology   ISBN: *1-59462-905-6*   *MSRP $15.45*

QTY

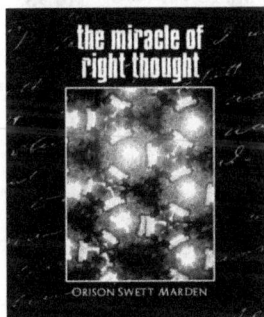

The Miracle of Right Thought
Orison Swett Marden

Believe with all of your heart that you will do what you were made to do. When the mind has once formed the habit of holding cheerful, happy, prosperous pictures, it will not be easy to form the opposite habit. It does not matter how improbable or how far away this realization may see, or how dark the prospects may be, if we visualize them as best we can, as vividly as possible, hold tenaciously to them and vigorously struggle to attain them, they will gradually become actualized, realized in the life. But a desire, a longing without endeavor, a yearning abandoned or held indifferently will vanish without realization.

Pages:360

Self Help   ISBN: *1-59462-644-8*   *MSRP $25.45*

The Rosicrucian Cosmo-Conception Mystic Christianity by *Max Heindel*   ISBN: *1-59462-188-8* **$38.95**
The Rosicrucian Cosmo-conception is not dogmatic, neither does it appeal to any other authority than the reason of the student. It is: not controversial, but is: sent forth in the, hope that it may help to clear...   New Age/Religion Pages 646

Abandonment To Divine Providence by *Jean-Pierre de Caussade*   ISBN: *1-59462-228-0* **$25.95**
"The Rev. Jean Pierre de Caussade was one of the most remarkable spiritual writers of the Society of Jesus in France in the 18th Century. His death took place at Toulouse in 1751. His works have gone through many editions and have been republished...   Inspirational/Religion Pages 400

Mental Chemistry by *Charles Haanel*   ISBN: *1-59462-192-6* **$23.95**
Mental Chemistry allows the change of material conditions by combining and appropriately utilizing the power of the mind. Much like applied chemistry creates something new and unique out of careful combinations of chemicals the mastery of mental chemistry...   New Age Pages 354

The Letters of Robert Browning and Elizabeth Barret Barrett 1845-1846 vol II   ISBN: *1-59462-193-4* **$35.95**
by *Robert Browning* and *Elizabeth Barrett*   Biographies Pages 596

Gleanings In Genesis (volume I) by *Arthur W. Pink*   ISBN: *1-59462-130-6* **$27.45**
Appropriately has Genesis been termed "the seed plot of the Bible" for in it we have, in germ form, almost all of the great doctrines which are afterwards fully developed in the books of Scripture which follow...   Religion/Inspirational Pages 420

The Master Key by *L. W. de Laurence*   ISBN: *1-59462-001-6* **$30.95**
In no branch of human knowledge has there been a more lively increase of the spirit of research during the past few years than in the study of Psychology, Concentration and Mental Discipline. The requests for authentic lessons in Thought Control, Mental Discipline and...   New Age/Business Pages 422

The Lesser Key Of Solomon Goetia by *L. W. de Laurence*   ISBN: *1-59462-092-X* **$9.95**
This translation of the first book of the "Lernegton" which is now for the first time made accessible to students of Talismanic Magic was done, after careful collation and edition, from numerous Ancient Manuscripts in Hebrew, Latin, and French...   New Age/Occult Pages 92

Rubaiyat Of Omar Khayyam by *Edward Fitzgerald*   ISBN:*1-59462-332-5* **$13.95**
Edward Fitzgerald, whom the world has already learned, in spite of his own efforts to remain within the shadow of anonymity, to look upon as one of the rarest poets of the century, was born at Bredfield, in Suffolk, on the 31st of March, 1809. He was the third son of John Purcell...   Music Pages 172

Ancient Law by *Henry Maine*   ISBN: *1-59462-128-4* **$29.95**
The chief object of the following pages is to indicate some of the earliest ideas of mankind, as they are reflected in Ancient Law, and to point out the relation of those ideas to modern thought.   Religion/History Pages 452

Far-Away Stories by *William J. Locke*   ISBN: *1-59462-129-2* **$19.45**
"Good wine needs no bush, but a collection of mixed vintages does. And this book is just such a collection. Some of the stories I do not want to remain buried for ever in the museum files of dead magazine-numbers an author's not unpardonable vanity..."   Fiction Pages 272

Life of David Crockett by *David Crockett*   ISBN: *1-59462-250-7* **$27.45**
"Colonel David Crockett was one of the most remarkable men of the times in which he lived. Born in humble life, but gifted with a strong will, an indomitable courage, and unremitting perseverance...   Biographies/New Age Pages 424

Lip-Reading by *Edward Nitchie*   ISBN: *1-59462-206-X* **$25.95**
Edward B. Nitchie, founder of the New York School for the Hard of Hearing, now the Nitchie School of Lip-Reading, Inc, wrote "LIP-READING Principles and Practice". The development and perfecting of this meritorious work on lip-reading was an undertaking...   How-to Pages 400

A Handbook of Suggestive Therapeutics, Applied Hypnotism, Psychic Science   ISBN: *1-59462-214-0* **$24.95**
by *Henry Munro*   Health/New Age/Health/Self-help Pages 376

A Doll's House: and Two Other Plays by *Henrik Ibsen*   ISBN: *1-59462-112-8* **$19.95**
Henrik Ibsen created this classic when in revolutionary 1848 Rome. Introducing some striking concepts in playwriting for the realist genre, this play has been studied the world over.   Fiction/Classics/Plays 308

The Light of Asia by *sir Edwin Arnold*   ISBN: *1-59462-204-3* **$13.95**
In this poetic masterpiece, Edwin Arnold describes the life and teachings of Buddha. The man who was to become known as Buddha to the world was born as Prince Gautama of India but he rejected the worldly riches and abandoned the reigns of power when...   Religion/History/Biographies Pages 170

The Complete Works of Guy de Maupassant by *Guy de Maupassant*   ISBN: *1-59462-157-8* **$16.95**
"For days and days, nights and nights, I had dreamed of that first kiss which was to consecrate our engagement, and I knew not on what spot I should put my lips..."   Fiction/Classics Pages 240

The Art of Cross-Examination by *Francis L. Wellman*   ISBN: *1-59462-309-0* **$26.95**
Written by a renowned trial lawyer, Wellman imparts his experience and uses case studies to explain how to use psychology to extract desired information through questioning.   How-to/Science/Reference Pages 408

Answered or Unanswered? by *Louisa Vaughan*   ISBN: *1-59462-248-5* **$10.95**
Miracles of Faith in China   Religion Pages 112

The Edinburgh Lectures on Mental Science (1909) by *Thomas*   ISBN: *1-59462-008-3* **$11.95**
This book contains the substance of a course of lectures recently given by the writer in the Queen Street Hall, Edinburgh. Its purpose is to indicate the Natural Principles governing the relation between Mental Action and Material Conditions...   New Age/Psychology Pages 148

Ayesha by *H. Rider Haggard*   ISBN: *1-59462-301-5* **$24.95**
Verily and indeed it is the unexpected that happens! Probably if there was one person upon the earth from whom the Editor of this, and of a certain previous history, did not expect to hear again...   Classics Pages 380

Ayala's Angel by *Anthony Trollope*   ISBN: *1-59462-352-X* **$29.95**
The two girls were both pretty, but Lucy who was twenty-one who supposed to be simple and comparatively unattractive, whereas Ayala was credited, as her Bombwhat romantic name might show, with poetic charm and a taste for romance. Ayala when her father died was nineteen...   Fiction Pages 484

The American Commonwealth by *James Bryce*   ISBN: *1-59462-286-8* **$34.45**
An interpretation of American democratic political theory. It examines political mechanics and society from the perspective of Scotsman James Bryce   Politics Pages 572

Stories of the Pilgrims by *Margaret P. Pumphrey*   ISBN: *1-59462-116-0* **$17.95**
This book explores pilgrims religious oppression in England as well as their escape to Holland and eventual crossing to America on the Mayflower, and their early days in New England...   History Pages 268

QTY

The Fasting Cure *by Sinclair Upton*   ISBN: *1-59462-222-1*   **$13.95**
In the Cosmopolitan Magazine for May, 1910, and in the Contemporary Review (London) for April, 1910, I published an article dealing with my experiences in fasting. I have written a great many magazine articles, but never one which attracted so much attention...   New Age/Self Help/Health Pages 164

Hebrew Astrology *by Sepharial*   ISBN: *1-59462-308-2*   **$13.45**
In these days of advanced thinking it is a matter of common observation that we have left many of the old landmarks behind and that we are now pressing forward to greater heights and to a wider horizon than that which represented the mind-content of our progenitors...   Astrology Pages 144

Thought Vibration or The Law of Attraction in the Thought World   ISBN: *1-59462-127-6*   **$12.95**
by William Walker Atkinson   Psychology/Religion Pages 144

Optimism *by Helen Keller*   ISBN: *1-59462-108-X*   **$15.95**
Helen Keller was blind, deaf, and mute since 19 months old, yet famously learned how to overcome these handicaps, communicate with the world, and spread her lectures promoting optimism. An inspiring read for everyone...   Biographies/Inspirational Pages 84

Sara Crewe *by Frances Burnett*   ISBN: *1-59462-360-0*   **$9.45**
In the first place, Miss Minchin lived in London. Her home was a large, dull, tall one, in a large, dull square, where all the houses were alike, and all the sparrows were alike, and where all the door-knockers made the same heavy sound...   Childrens/Classic Pages 88

The Autobiography of Benjamin Franklin *by Benjamin Franklin*   ISBN: *1-59462-135-7*   **$24.95**
The Autobiography of Benjamin Franklin has probably been more extensively read than any other American historical work, and no other book of its kind has had such ups and downs of fortune. Franklin lived for many years in England, where he was agent...   Biographies/History Pages 332

Name	
Email	
Telephone	
Address	
City, State ZIP	

☐ **Credit Card**   ☐ **Check / Money Order**

Credit Card Number	
Expiration Date	
Signature	

Please Mail to: Book Jungle
PO Box 2226
Champaign, IL 61825
or Fax to: 630-214-0564

ORDERING INFORMATION

web: *www.bookjungle.com*
email: *sales@bookjungle.com*
fax: *630-214-0564*
mail: *Book Jungle PO Box 2226 Champaign, IL 61825*
or PayPal *to sales@bookjungle.com*

Please contact us for bulk discounts

DIRECT-ORDER TERMS

20% Discount if You Order
Two or More Books
Free Domestic Shipping!
Accepted: Master Card, Visa,
Discover, American Express